# CR SUBMANIFOLDS
# OF COMPLEX PROJECTIVE SPACE

# Developments in Mathematics

## VOLUME 19

*Series Editor:*
Krishnaswami Alladi, *University of Florida, U.S.A.*

# CR SUBMANIFOLDS
# OF COMPLEX PROJECTIVE SPACE

By

MIRJANA DJORIĆ
University of Belgrade, Serbia

MASAFUMI OKUMURA
Saitama University, Japan

 Springer

Mirjana Djorić
Faculty of Mathematics
University of Belgrade
11000, Belgrade
Serbia
mdjoric@matf.bg.ac.rs

Masafumi Okumura
Professor Emeritus
Saitama University
Saitama, 338-8570
Japan
mokumura@h8.dion.ne.jp

ISSN 1389-2177
ISBN 978-1-4614-2477-2          e-ISBN 978-1-4419-0434-8
DOI 10.1007/978-1-4419-0434-8
Springer New York Dordrecht Heidelberg London

Mathematics Subject Classification (2000): 53C15, 53C40, 53B20, 53B25, 53B35, 53C20, 53C25, 53C42, 53C55, 53D15

Springer is part of Springer Science+Business Media (www.springer.com)

# Contents

# Preface

Although submanifolds complex manifolds has been an active field of study for many years, in some sense this area is not sufficiently covered in the current literature. This text deals with the CR submanifolds of complex manifolds, with particular emphasis on CR submanifolds of complex projective space, and it covers the topics which are necessary for learning the basic properties of these manifolds. We are aware that it is impossible to give a complete overview of these submanifolds, but we hope that these notes can serve as an introduction to their study. We present the fundamental definitions and results necessary for reaching the frontiers of research in this field.

There are many monographs dealing with some current interesting topics in differential geometry, but most of these are written as encyclopedias, or research monographs, gathering recent results and giving the readers ample useful information about the topics. Therefore, these kinds of monographs are attractive to specialists in differential geometry and related fields and acceptable to professional differential geometers. However, for graduate students who are less advanced in differential geometry, these texts might be hard to read without assistance from their instructors. By contrast, the general philosophy of this book is to begin with the elementary facts about complex manifolds and their submanifolds, give some details and proofs, and introduce the reader to the study of CR submanifolds of complex manifolds; especially complex projective space. It includes only a few original results with precise proofs, while the others are cited in the reference list. For this reason this book is appropriate for graduate students majoring in differential geometry and for researchers who are interested in geometry of complex manifolds and its submanifolds.

Additionally, this research monograph is intended to give a rapid and accessible introduction to particular subjects, guiding the audience to topics of current research and to more advanced and specialized literature, collecting many results previously available only in research papers and providing references to many other recently published papers. Our aim has been to give a reasonably comprehensive and self-contained account of the subject, presenting mathematical results that are new or have not previously been accessible in the literature. Our intention has been not only to provide relevant techniques, results and their applications, but also to afford insight into the motivations and ideas behind the theory.

The prerequisites for this text are the knowledge of the introductory manifold theory and of curvature properties of Riemannian geometry. Although we intended to write this material to be self-contained, as much as possible, and to give complete proofs of all standard results, some basic results could not be written only with the basic knowledge of Riemannian geometry and for these results we only cite the references.

The first half of the text covers the basic material about the geometry of submanifolds of complex manifolds. Special topics that are explored include the (almost) complex structure, Kähler manifold, submersion and immersion, and the structure equations of a submanifold. This part is based on the second author's lectures, given at Saitama University, Japan.

The second part of the text deals with real hypersurfaces and CR submanifolds, with particular emphasis on CR submanifolds of maximal CR dimension in complex projective space. Fundamental results which are not new, but recently published in some mathematical journals, are presented in detail. The final six chapters contain the original results by the authors with complete proofs.

We would like to express our appreciation to D. Blair, P. Bueken, A. Hinić, M. Lukić, S. Nagai, M. Prvanović, L. Vanhecke, who spent considerable time and effort in reading the original notes and who supplied us with valuable suggestions, which resulted in many improvements of both the content and the presentation.

We would also like to thank E. Loew of Springer and J. L. Spiegelman for their kind assistance in the production of this book.

Mirjana Djorić<br>
Masafumi Okumura<br>
June, 2009

# 1

# Complex manifolds

Let us first recall the definition of a holomorphic function. Denote by $\mathbf{C}$ the field of complex numbers. For a positive integer $n$, the $n$-dimensional complex number space

$$\mathbf{C}^n = \{z \,|\, z = (z^1, \ldots, z^n), \ z^j \in \mathbf{C} \quad \text{for} \quad 1 \le j \le n\}$$

is the Cartesian product of $n$ copies of $\mathbf{C}$. The standard *Hermitian inner product* on $\mathbf{C}^n$ is defined by

$$(a, b) = \sum_{j=1}^{n} a^j \overline{b}^j, \quad a, b \in \mathbf{C}^n.$$

The associated *norm* $|a| = (a, a)^{\frac{1}{2}}$ induces the Euclidean metric in the usual way: for $a, b \in \mathbf{C}^n$, $dist(a, b) = |a - b|$.

The (open) ball of radius $r > 0$ and center $a \in \mathbf{C}^n$ is defined by

$$B(a, r) = \{z \in \mathbf{C}^n \,|\, |z - a| < r\}.$$

The collection of balls $\{B(a, r) : r > 0 \quad \text{and rational}\}$ forms a countable neighborhood basis at the point $a$ for the topology of $\mathbf{C}^n$. The topology of $\mathbf{C}^n$ is identical with the one arising from the following identification (which will be used throughout this manuscript) of $\mathbf{C}^n$ with $\mathbf{R}^{2n}$, where

$$\mathbf{R}^{2n} = \{(x^1, \ldots, x^{2n}), \ x^j \in \mathbf{R} \quad \text{for} \quad 1 \le j \le 2n\}.$$

Given $z = (z^1, \ldots, z^n) \in \mathbf{C}^n$, each coordinate $z^j$ can be written as $z^j = x^j + \sqrt{-1}y^j$, with $x^j, y^j \in \mathbf{R}$. The mapping

$$\mathbf{C}^n \ni z \mapsto (x^1, y^1, \ldots, x^n, y^n) \in \mathbf{R}^{2n}$$

establishes an $\mathbf{R}$-linear isomorphism between $\mathbf{C}^n$ and $\mathbf{R}^{2n}$, which is compatible with the metric structures: a ball $B(a, r)$ in $\mathbf{C}^n$ is identified with a

M. Djorić, M. Okumura, *CR Submanifolds of Complex Projective Space*,
Developments in Mathematics 19, DOI 10.1007/978-1-4419-0434-8_1,
© Springer Science+Business Media, LLC 2010

Euclidean ball in $\mathbf{R}^{2n}$ of equal radius $r$. Because of this identification, all the usual concepts from topology and analysis on real Euclidean spaces $\mathbf{R}^{2n}$ carry over immediately to $\mathbf{C}^n$. In particular, we recall that $D \subset \mathbf{C}^n$ is open if for every $a \in D$ there is a ball $B(a, r) \subset D$ with $r > 0$, and that an open set $D \subset \mathbf{C}^n$ is connected if and only if $D$ is pathwise connected.

We now introduce the class of functions which is the principal object in this section.

**Definition 1.1.** Let $D$ be an open subset of $\mathbf{C}^n$. A function $f : D \to \mathbf{C}$ is called *differentiable* at $z_0$, if

$$\lim_{h \to 0} \frac{1}{h} \left\{ f(z_0^1, \ldots, z_0^i + h, \ldots, z_0^n) - f(z_0^1, \ldots, z_0^i, \ldots, z_0^n) \right\}$$

exists for every $i = 1, \ldots, n$. $f$ is called *holomorphic* on $D$ if $f$ is differentiable at any point of $D$.

If we denote this limit by $c^i$, the above condition is equivalent to

$$f(z_0^1, \ldots, z_0^i + h, \ldots, z_0^n) - f(z_0^1, \ldots, z_0^i, \ldots, z_0^n) - hc^i = \alpha^i(h)|h| \qquad (1.1)$$

where $h \to 0$ implies $\alpha^i(h) \to 0$.

We put $z^i = x^i + \sqrt{-1}y^i$ and $h = t + \sqrt{-1}s$. Then $(z^1, \ldots, z^n) \in \mathbf{C}^n$ is identified with $(x^1, y^1, \ldots, x^n, y^n) \in \mathbf{R}^{2n}$ and consequently, (1.1) is equivalent to

$$f(\ldots, x_0^i + t, y_0^i + s, \ldots) - f(\ldots, x_0^i, y_0^i, \ldots) = c^i(t + \sqrt{-1}s) + \alpha^i(t, s)|h|,$$

for $|h| = \sqrt{t^2 + s^2}$. With the notation

$$c^i = a^i + \sqrt{-1}\,b^i, \quad \alpha^i = \beta^i + \sqrt{-1}\,\gamma^i, \quad f = u + \sqrt{-1}\,v,$$

we compute

$$\begin{aligned}
u(\ldots, x_0^i + t, y_0^i &+ s, \ldots) + \sqrt{-1}v(\ldots, x_0^i + t, y_0^i + s, \ldots) \\
-u(\ldots, x_0^i, y_0^i, \ldots) &- \sqrt{-1}v(\ldots, x_0^i, y_0^i, \ldots) \\
&= (a^i t - b^i s) + \sqrt{-1}(a^i s + b^i t) + \sqrt{t^2 + s^2}(\beta^i + \sqrt{-1}\gamma^i),
\end{aligned}$$

that is,

$$\begin{aligned}
u(\ldots, x_0^i + t, y_0^i + s, \ldots) - u(\ldots, x_0^i, y_0^i, \ldots) &= a^i t - b^i s + \sqrt{t^2 + s^2}\beta^i, \\
v(\ldots, x_0^i + t, y_0^i + s, \ldots) - v(\ldots, x_0^i, y_0^i, \ldots) &= a^i s + b^i t + \sqrt{t^2 + s^2}\gamma^i.
\end{aligned}$$

Thus,

$$h \to 0 \quad \text{implies} \quad \alpha(h) \to 0$$

is equivalent to requiring that

$$t \to 0, s \to 0 \quad \text{implies} \quad \beta \to 0, \gamma \to 0.$$

This shows that the real functions $u$ and $v$ are both totally differentiable at $z_0$.

Taking the limit along the real axis, that is, $h = t \to 0$, of a holomorphic function $f = u + \sqrt{-1}v$, we compute

$$\frac{\partial u}{\partial x^i}(\ldots, x_0^i, y_0^i, \ldots) = a^i, \qquad \frac{\partial v}{\partial x^i}(\ldots, x_0^i, y_0^i, \ldots) = b^i.$$

In the same way, taking the limit along the imaginary axis, that is, $h = s \to 0$, we obtain

$$\frac{\partial u}{\partial y^i}(\ldots, x_0^i, y_0^i, \ldots) = -b^i, \qquad \frac{\partial v}{\partial y^i}(\ldots, x_0^i, y_0^i, \ldots) = a^i.$$

Therefore, we conclude that if $f = u + \sqrt{-1}v$ is a holomorphic function, then the real functions $u$ and $v$ satisfy the following Cauchy-Riemann equations:

$$\frac{\partial u}{\partial x^i} = \frac{\partial v}{\partial y^i}, \qquad \frac{\partial u}{\partial y^i} = -\frac{\partial v}{\partial x^i}. \tag{1.2}$$

Now, we consider the converse. Let $u$ and $v$ be differentiable functions that satisfy the Cauchy-Riemann equations (1.2) and let $f = u + \sqrt{-1}v$. Then

$$f(z_0^1, \ldots, z_0^i + h, \ldots, z_0^n) - f(z_0^1, \ldots, z_0^i, \ldots, z_0^n) = u(\ldots, x_0^i + t, y_0^i + s, \ldots)$$
$$-u(\ldots, x_0^i, y_0^i, \ldots) - \sqrt{-1}\left\{v(\ldots, x_0^i + t, y_0^i + s, \ldots) - v(\ldots, x_0^i, y_0^i, \ldots)\right\}.$$

Using the mean value theorem, we compute

$$u(\ldots, x_0^i + t, y_0^i + s, \ldots) - u(\ldots, x_0^i, y_0^i, \ldots) = u(\ldots, x_0^i + t, y_0^i + s, \ldots)$$
$$-u(\ldots, x_0^i, y_0^i + s, \ldots) + u(\ldots, x_0^i, y_0^i + s, \ldots) - u(\ldots, x_0^i, y_0^i, \ldots)$$
$$= \frac{\partial u}{\partial x^i}\left(\ldots, x_0^i + \theta_1 t, y_0^i + s, \ldots\right) t + \frac{\partial u}{\partial y^i}\left(\ldots, x_0^i, y_0^i + \theta_2 s, \ldots\right) s$$
$$= \left(\frac{\partial u}{\partial x^i}(\ldots, x_0^i, y_0^i, \ldots) + \epsilon_1\right) t + \left(\frac{\partial u}{\partial y^i}(\ldots, x_0^i, y_0^i, \ldots) + \epsilon_2\right) s,$$

where $0 < \theta_1, \theta_2 < 1$ and $\epsilon_1, \epsilon_2 \to 0$ when $|h| \to 0$. Similarly, we have

$$v(\ldots, x_0^i + t, y_0^i + s, \ldots) - v(\ldots, x_0^i, y_0^i, \ldots)$$
$$= \left(\frac{\partial v}{\partial x^i}(\ldots, x_0^i, y_0^i, \ldots) + \epsilon_3\right) t + \left(\frac{\partial v}{\partial y^i}(\ldots, x_0^i, y_0^i, \ldots) + \epsilon_4\right) s.$$

Hence, using the Cauchy-Riemann equations, we obtain

$$\frac{1}{h}\left\{ f(z_0^1,\ldots,z_0+h,\ldots,z_0^n) - f(z_0^1,\ldots,z_0^i,\ldots,z_0^n) \right\}$$

$$= \left( \frac{\partial u}{\partial x^i}(\ldots,x_0^i,y_0^i,\ldots) + \epsilon_1 \right) \frac{t}{h} + \left( \frac{\partial u}{\partial y^i}(\ldots,x_0^i,y_0^i,\ldots) + \epsilon_2 \right) \frac{s}{h}$$

$$+ \sqrt{-1}\left\{ \left( \frac{\partial v}{\partial x^i}(\ldots,x_0^i,y_0^i,\ldots) + \epsilon_3 \right) \frac{t}{h} + \left( \frac{\partial v}{\partial y^i}(\ldots,x_0^i,y_0^i,\ldots) + \epsilon_4 \right) \frac{s}{h} \right\}$$

$$= \left( \frac{\partial u}{\partial x^i}(\ldots,x_0^i,y_0^i,\ldots) + \epsilon_1 \right) \frac{t}{h} + \sqrt{-1}\left( \frac{\partial u}{\partial x^i}(\ldots,x_0^i,y_0^i,\ldots) + \epsilon_4 \right) \frac{s}{h}$$

$$+ \sqrt{-1}\left( \frac{\partial v}{\partial x^i}(\ldots,x_0^i,y_0^i,\ldots) + \epsilon_3 \right) \frac{t}{h} - \left( \frac{\partial v}{\partial x^i}(\ldots,x_0^i,y_0^i,\ldots) + \epsilon_2 \right) \frac{s}{h}$$

$$= \frac{1}{h}\left( \frac{\partial u}{\partial x_i}(\ldots,x_0^i,y_0^i,\ldots) + \sqrt{-1}\frac{\partial v}{\partial x^i}(\ldots,x_0^i,y_0^i,\ldots) \right)(t + \sqrt{-1}s)$$

$$+ \delta_1 \frac{t}{h} + \delta_2 \frac{s}{h}$$

$$= \frac{\partial u}{\partial x_i}(\ldots,x_0^i,y_0^i,\ldots) + \sqrt{-1}\frac{\partial v}{\partial x^i}(\ldots,x_0^i,y_0^i,\ldots) + \delta_1 \frac{t}{h} + \delta_2 \frac{s}{h},$$

where $\delta_1, \delta_2 \to 0$ whenever $h \to 0$. Since $|t/h| \leq 1$, $|s/h| \leq 1$, we conclude

$$\lim_{h\to 0} \frac{1}{h}\left\{ f(z_0^1,\ldots,z_0^i+h,\ldots,z_0^n) - f(z_0^1,\ldots,z_0^i,\ldots,z_0^n) \right\}$$

$$= \frac{\partial u}{\partial x^i}(\ldots,x_0^i,y_0^i,\ldots) + \sqrt{-1}\frac{\partial v}{\partial x^i}(\ldots,x_0^i,y_0^i,\ldots).$$

Thus, if differentiable real functions $u$ and $v$ satisfy the Cauchy-Riemann equations, then $f = u + \sqrt{-1}v$ is differentiable.

**Definition 1.2.** Let $D$ be an open subset of $\mathbf{C}^n$ and let $\psi$ be a mapping: $D \to \mathbf{C}^n$ defined by

$$\psi(z^1,\ldots,z^n) = (w^1,\ldots,w^n).$$

$\psi$ is *holomorphic* if, for each $i$, functions $w^i = \psi^i(z^1,\ldots,z^n)$ are holomorphic with respect to $z^j$, $j = 1,\ldots,n$.

Now we recall the definition of a complex manifold. Roughly speaking, a complex manifold is a topological space that locally looks like a neighborhood in $\mathbf{C}^n$. To be precise, we have

**Definition 1.3.** A Hausdorff space $M$ is called a *complex manifold* of (complex) dimension $n$, if $M$ satisfies the following properties:

(1) there exists an open covering $\{U_\alpha\}_{\alpha \in A}$ of $M$ and, for each $\alpha$, there exists a homeomorphism

$$\psi_\alpha : U_\alpha \to \psi_\alpha(U_\alpha) \subset \mathbf{C}^n;$$

(2) for any two open sets $U_\alpha$ and $U_\beta$ with nonempty intersection, maps

$$f_{\beta\alpha} = \psi_\beta \circ \psi_\alpha^{-1} : \psi_\alpha(U_\alpha \cap U_\beta) \to \psi_\beta(U_\alpha \cap U_\beta),$$
$$f_{\alpha\beta} = \psi_\alpha \circ \psi_\beta^{-1} : \psi_\beta(U_\alpha \cap U_\beta) \to \psi_\alpha(U_\alpha \cap U_\beta)$$

are holomorphic.

The set $\{(U_\alpha, \psi_\alpha)\}_{\alpha \in A}$ is called a *system of holomorphic coordinate neighborhoods*.

We will often use the superscript to denote the dimension of a manifold. The symbol $M^n$ means that $M$ is a manifold of (complex) dimension $n$.

Next, we consider some examples of complex manifolds. From the definition, it is clear that the product of two complex manifolds, or a connected open subset in a complex manifold, are complex manifolds.

*Example* 1.1. An $n$-dimensional complex space $\mathbf{C}^n$ and an open set of $\mathbf{C}^n$ are complex manifolds. We may take the identity map *id* for $\psi$.   $\diamondsuit$

*Example* 1.2. *Riemann sphere.*

Let
$$\mathbf{S}^2 = \left\{ (x, y, z) \in \mathbf{R}^3 \,|\, x^2 + y^2 + z^2 = 1 \right\}.$$

We put $U_1 = \mathbf{S}^2 \backslash \{n\}$ and $U_2 = \mathbf{S}^2 \backslash \{s\}$, where $n = (0, 0, 1)$ and $s = (0, 0, -1)$. We define $\psi_1 : U_1 \to \mathbf{C}$ and $\psi_2 : U_2 \to \mathbf{C}$ to be the stereographic maps from $n$ and $s$, respectively, that is,

$$\psi_1(x, y, z) = \frac{x + \sqrt{-1}\,y}{1 - z}, \qquad \psi_2(x, y, z) = \frac{x - \sqrt{-1}\,y}{1 + z}.$$

Then maps $\psi_1 \circ \psi_2^{-1}$, $\psi_2 \circ \psi_1^{-1} : \mathbf{C} \to \mathbf{C}$ are holomorphic. Namely, for $\mathbf{C} \ni w = u + \sqrt{-1}v$, from $u = \frac{x}{1-z}$, $v = \frac{y}{1-z}$ and $x^2 + y^2 + z^2 = 1$, we conclude

$$z = \frac{u^2 + v^2 - 1}{u^2 + v^2 + 1}, \quad x = \frac{2u}{u^2 + v^2 + 1}, \quad y = \frac{2v}{u^2 + v^2 + 1}.$$

Hence

$$\psi_1^{-1}(w) = \psi_1^{-1}(u + \sqrt{-1}v) = \left( \frac{2u}{u^2 + v^2 + 1}, \frac{2v}{u^2 + v^2 + 1}, \frac{u^2 + v^2 - 1}{u^2 + v^2 + 1} \right)$$

from which we have

$$\psi_2 \circ \psi_1^{-1}(w) = \frac{u}{u^2 + v^2} - \sqrt{-1}\frac{v}{u^2 + v^2} = \frac{1}{w}.$$

Thus $\psi_2 \circ \psi_1^{-1}$ is holomorphic. Similarly we can also prove that $\psi_1 \circ \psi_2^{-1}$ is holomorphic. Therefore, $\mathbf{S}^2$ is a complex manifold, called Riemann sphere.   $\diamondsuit$

*Example* 1.3. *Complex projective space* $\mathbf{P}^n(\mathbf{C})$.

Let $z = (z^1, \ldots, z^{n+1})$ and $w = (w^1, \ldots, w^{n+1}) \in \mathbf{C}^{n+1} \setminus \{0\}$ and set $w \sim z$, if there exists a non-zero complex number $\alpha$ such that $w = \alpha z$. Then $\sim$ defines the equivalence relation in $\mathbf{C}^{n+1} \setminus \{0\}$. The *complex projective space* $\mathbf{P}^n(\mathbf{C})$ is the set of equivalence classes $\mathbf{C}^{n+1} \setminus \{0\}/\sim$ with the quotient topology from $\mathbf{C}^{n+1} \setminus \{0\}$.

Denote

$$U_\alpha = \{[(z^1, \ldots, z^\alpha, \ldots, z^{n+1})] \in \mathbf{P}^n(\mathbf{C}) \,|\, z^\alpha \neq 0\}$$

and let $\psi_\alpha : U_\alpha \to \mathbf{C}^n$ be the map defined by

$$\psi_\alpha \left([(z^1, \ldots, z^\alpha, \ldots, z^{n+1})]\right) = \left(\frac{z^1}{z^\alpha}, \ldots, \frac{z^{\alpha-1}}{z^\alpha}, \frac{z^{\alpha+1}}{z^\alpha}, \ldots, \frac{z^{n+1}}{z^\alpha}\right).$$

Then, $\psi_\alpha^{-1}(w^1, \ldots, w^n) = [(w^1, \ldots, w^{\alpha-1}, 1, w^\alpha, \ldots, w^n)]$ and therefore

$$\psi_\beta \circ \psi_\alpha^{-1}(z^1, \ldots, z^n) = \left(\frac{z^1}{z^\beta}, \ldots, \frac{z^{\alpha-1}}{z^\beta}, \frac{1}{z^\beta}, \frac{z^\alpha}{z^\beta}, \ldots, \frac{z^{\beta-1}}{z^\beta}, \frac{z^{\beta+1}}{z^\beta}, \ldots, \frac{z^n}{z^\beta}\right).$$

Thus, $\psi_\beta \circ \psi_\alpha^{-1}$ is holomorphic and the complex projective space is a complex manifold.     $\diamond$

**Definition 1.4.** Let $(U, \psi)$ be a holomorphic coordinate neighborhood of a complex manifold $M$. A function $f : U \to \mathbf{C}$ is *holomorphic* if the function $f \circ \psi^{-1} : \psi(U) \to \mathbf{C}$ is holomorphic.

**Definition 1.5.** Let $M$, $N$ be complex manifolds and $(U, \psi)$ a holomorphic coordinate neighborhood of $x \in M$. A continuous map $\phi : M \to N$ is *holomorphic* if for any $x \in M$ and for any holomorphic coordinate neighborhood $(V, \psi')$ of $N$ such that $\phi(x) \in V$ and $\phi(U) \subset V$, $\psi' \circ \phi \circ \psi^{-1} : \psi(U) \to \psi'(V)$ is holomorphic.

Since the coordinate changes are biholomorphic (i.e., two-way holomorphic), the above definition of holomorphicity for maps is independent of the choice of local holomorphic neighborhood systems.

**Definition 1.6.** $M$ is called a *complex submanifold* of a complex manifold $\overline{M}$, if $M$ satisfies the following conditions:

(1) $M$ is a submanifold of $\overline{M}$ as a differentiable manifold;

(2) the injection $\imath : M \to \overline{M}$ is holomorphic.

# 2

# Almost complex structure

We recall the definition of an almost complex structure. First, we identify a complex number $z = x + \sqrt{-1}\,y$ with the element $z = x\,e_1 + y\,e_2$ of a two-dimensional vector space $V$, where $(e_1, e_2)$ denotes the basis of $V$. Let $I : V \to V$ be the endomorphism defined by

$$Iz = \sqrt{-1}z = -y + ix.$$

Then we conclude

$$xIe_1 + yIe_2 = I(xe_1 + ye_2) = Iz = -ye_1 + xe_2.$$

Therefore, the endomorphism $I$ is determined by

$$Ie_1 = e_2, \quad Ie_2 = -e_1.$$

Keeping this in mind, we introduce the endomorphism $J$ of the tangent space $T_x(M)$ of a complex manifold $M$ at $x \in M$.

Let $M$ be an $n$-dimensional complex manifold. Identifying the local complex coordinates $(z^1, \ldots, z^n)$ with $(x^1, y^1, \ldots, x^n, y^n)$, where $z^i = x^i + \sqrt{-1}y^i$, $i = 1, \ldots, n$, we regard $M$ as a $2n$-dimensional differentiable manifold. The tangent space $T_x(M)$ of $M$ at a point $x \in M$ has a natural basis $\left\{ (\frac{\partial}{\partial x^1})_x, (\frac{\partial}{\partial y^1})_x, \ldots, (\frac{\partial}{\partial x^n})_x, (\frac{\partial}{\partial y^n})_x \right\}$. For $i = 1, \ldots, n$, we put

$$J_x \left( \frac{\partial}{\partial x^i} \right)_x = \left( \frac{\partial}{\partial y^i} \right)_x, \quad J_x \left( \frac{\partial}{\partial y^i} \right)_x = - \left( \frac{\partial}{\partial x^i} \right)_x. \qquad (2.1)$$

Then $J_x$ defines an isomorphism $J_x : T_x(M) \to T_x(M)$. In fact, if we take other local complex coordinates $(w^1, \ldots, w^n)$, where $w^i = u^i + \sqrt{-1}v^i$, then they satisfy the Cauchy-Riemann equations,

$$\frac{\partial x^i}{\partial u^j} = \frac{\partial y^i}{\partial v^j}, \quad \frac{\partial x^i}{\partial v^j} = -\frac{\partial y^i}{\partial u^j}$$

M. Djorić, M. Okumura, *CR Submanifolds of Complex Projective Space*, Developments in Mathematics 19, DOI 10.1007/978-1-4419-0434-8_2, © Springer Science+Business Media, LLC 2010

for $i, j = 1, \ldots, n$. Hence

$$
\begin{aligned}
J_x \left( \frac{\partial}{\partial u^i} \right)_x &= \sum_j \left( \frac{\partial x^j}{\partial u^i} J_x \left( \frac{\partial}{\partial x^j} \right)_x + \frac{\partial y^j}{\partial u^i} J_x \left( \frac{\partial}{\partial y^j} \right)_x \right) \\
&= \sum_j \left( \frac{\partial x^j}{\partial u^i} \left( \frac{\partial}{\partial y^j} \right)_x - \frac{\partial y^j}{\partial u^i} \left( \frac{\partial}{\partial x^j} \right)_x \right) \\
&= \sum_j \left( \frac{\partial y^j}{\partial v^i} \left( \frac{\partial}{\partial y^j} \right)_x + \frac{\partial x^j}{\partial v^i} \left( \frac{\partial}{\partial x^j} \right)_x \right) = \left( \frac{\partial}{\partial v^i} \right)_x,
\end{aligned}
$$

and

$$
\begin{aligned}
J_x \left( \frac{\partial}{\partial v^i} \right)_x &= \sum_j \left( \frac{\partial x^j}{\partial v^i} J_x \left( \frac{\partial}{\partial x^j} \right)_x + \frac{\partial y^j}{\partial v^i} J_x \left( \frac{\partial}{\partial y^j} \right)_x \right) \\
&= \sum_j \left( \frac{\partial x^j}{\partial v^i} \left( \frac{\partial}{\partial y^j} \right)_x - \frac{\partial y^j}{\partial v^i} \left( \frac{\partial}{\partial x^j} \right)_x \right) \\
&= \sum_j \left( -\frac{\partial y^j}{\partial u^i} \left( \frac{\partial}{\partial y^j} \right)_x - \frac{\partial x^j}{\partial u^i} \left( \frac{\partial}{\partial x^j} \right)_x \right) = - \left( \frac{\partial}{\partial u^i} \right)_x.
\end{aligned}
$$

Thus $J_x$ is independent of the choice of holomorphic coordinates and is well-defined. Regarding $J$ as a map of the *tangent bundle* $T(M) = \bigcup_{x \in M} T_x(M)$, we call $J$ the (natural) *almost complex structure* of $M$.

**Proposition 2.1.** *Let $M$ and $M'$ be complex manifolds with almost complex structures $J$ and $J'$, respectively. Then the map $\phi : M \to M'$ is holomorphic if and only if $\phi_* \circ J = J' \circ \phi_*$, where $\phi_*$ denotes the differential map of $\phi$.*

*Proof.* We identify holomorphic coordinates $(z^1, \ldots, z^n)$ of $M$ with $(x^1, y^1, \ldots, x^n, y^n)$ and holomorphic coordinates $(w^1, \ldots, w^m)$ of $M'$ with $(u^1, v^1, \ldots, u^m, v^m)$, where $z^i = x^i + \sqrt{-1} y^i$ and $w^j = u^j + \sqrt{-1} v^j$. Then

$$
\phi(z^1, \ldots, z^n) = (w^1(z^1, \ldots, z^n), \ldots, w^m(z^1, \ldots, z^n))
$$

is expressed by

$$
\begin{aligned}
\phi(\ldots, x^i, y^i, \ldots) &= (u^1(\ldots, x^i, y^i, \ldots), \\
v^1(\ldots, x^i, y^i, \ldots), \ldots, u^m(\ldots, x^i, y^i, \ldots), v^m(\ldots, x^i, y^i, \ldots))
\end{aligned}
$$

in terms of the real coordinates. Thus we have

$$J' \circ \phi_* \left( \frac{\partial}{\partial x^i} \right) = \sum_{j=1}^{m} \left( \frac{\partial u^j}{\partial x^i} J' \left( \frac{\partial}{\partial u^j} \right) + \frac{\partial v^j}{\partial x^i} J' \left( \frac{\partial}{\partial v^j} \right) \right) \qquad (2.2)$$

$$= \sum_{j=1}^{m} \left( \frac{\partial u^j}{\partial x^i} \frac{\partial}{\partial v^j} - \frac{\partial v^j}{\partial x^i} \frac{\partial}{\partial u^j} \right),$$

$$J' \circ \phi_* \left( \frac{\partial}{\partial y^i} \right) = \sum_{j=1}^{m} \left( \frac{\partial u^j}{\partial y^i} J' \left( \frac{\partial}{\partial u^j} \right) + \frac{\partial v^j}{\partial y^i} J' \left( \frac{\partial}{\partial v^j} \right) \right) \qquad (2.3)$$

$$= \sum_{j=1}^{m} \left( \frac{\partial u^j}{\partial y^i} \frac{\partial}{\partial v^j} - \frac{\partial v^j}{\partial y^i} \frac{\partial}{\partial u^j} \right).$$

On the other hand,

$$\phi_* \circ J \left( \frac{\partial}{\partial x^i} \right) = \phi_* \left( \frac{\partial}{\partial y^i} \right) = \sum_{j=1}^{m} \left( \frac{\partial u^j}{\partial y^i} \frac{\partial}{\partial u^j} + \frac{\partial v^j}{\partial y^i} \frac{\partial}{\partial v^j} \right), \qquad (2.4)$$

$$\phi_* \circ J \left( \frac{\partial}{\partial y^i} \right) = -\phi_* \left( \frac{\partial}{\partial x^i} \right) = -\sum_{j=1}^{m} \left( \frac{\partial u^j}{\partial x^i} \frac{\partial}{\partial u^j} + \frac{\partial v^j}{\partial x^i} \frac{\partial}{\partial v^j} \right). \qquad (2.5)$$

Comparing (2.2), (2.3) with (2.4), (2.5), yields the Cauchy-Riemann equations

$$\frac{\partial u^j}{\partial x^i} = \frac{\partial v^j}{\partial y^i}, \quad \frac{\partial u^j}{\partial y^i} = -\frac{\partial v^j}{\partial x^i}.$$

Consequently, $\phi$ is holomorphic if and only if $\phi_* \circ J = J' \circ \phi_*$. $\qquad \square$

**Definition 2.1.** A differentiable manifold $M$ is said to be an *almost complex manifold* if there exists a linear map $J : T(M) \to T(M)$ satisfying $J^2 = -id$ and $J$ is said to be an *almost complex structure* of $M$.

As we have shown, a complex manifold $M$ admits a naturally induced almost complex structure from the complex structure, given by (2.1), and consequently $M$ is an almost complex manifold.

**Proposition 2.2.** *An almost complex manifold $M$ is even-dimensional.*

*Proof.* Since $J^2 = -id$, for suitable basis of the tangent bundle we have

$$J^2 = \begin{pmatrix} -1 & 0 & \cdots & 0 \\ 0 & -1 & \cdots & 0 \\ & & \cdots\cdots & \\ 0 & \cdots & \cdots & -1 \end{pmatrix}.$$

Hence, $(-1)^n = \det J^2 = (\det J)^2 \geq 0$. Thus, $n$ is even. $\qquad \square$

*Remark* 2.1. Here we note that an even-dimensional differentiable manifold does not necessarily admit an almost complex structure $J$. It is known, for example, that $\mathbf{S}^4$ does not possess an almost complex structure (see [54]).

The *Nijenhuis tensor* $N$ of an almost complex structure $J$ is defined by

$$N(X,Y) = J[X,Y] - [JX,Y] - [X,JY] - J[JX,JY] \tag{2.6}$$

for any $X,Y \in T(M)$ and its tensorial property is established by the following

**Proposition 2.3.** *For a function $f$ on $M$, we have $N(fX,Y) = fN(X,Y)$.*

*Proof.* We note that $[fX,Y] = f[X,Y] - (Yf)X$ and therefore

$$\begin{aligned}
N(fX,Y) &= J[fX,Y] - [JfX,Y] - [fX,JY] - J[JfX,JY] \\
&= J[fX,Y] - [fJX,Y] - [fX,JY] - J[fJX,JY] \\
&= J(f[X,Y] - (Yf)X) - f[JX,Y] + (Yf)JX - f[X,JY] \\
&\quad + ((JY)f)X - J(f[JX,JY] - ((JY)f)JX) \\
&= f(J[X,Y] - [JX,Y] - [X,JY] - J[JX,JY]) \\
&\quad + ((JY)f)X + ((JY)f)J^2X \\
&= fN(X,Y),
\end{aligned}$$

which establishes the formula.                                  □

**Theorem 2.1.** *Let $M$ be an almost complex manifold with almost complex structure $J$. There exists a complex structure on $M$ and $J$ is the almost complex structure which is induced from the complex structure on $M$ if and only if the Nijenhuis tensor $N$ vanishes identically.*

*Proof.* If $M$ is a complex manifold, from Proposition 2.3, together with the definition of $J$, the necessity of the theorem is rather trivial. To prove the sufficiency of the theorem, we should use a theory of PDE and therefore we omit it. (See [39] for a detailed proof.)                          □

**Proposition 2.4.** *Let $(M,J)$ be an almost complex manifold and suppose that on $M$ there exists an open covering $\mathcal{U} = \{U_\alpha\}$ which satisfies the following condition:*
*For each $U_\alpha \in \mathcal{U}$, there is a local coordinate system $(x^1, x^2, \ldots, x^{2n})$ such that, at any point $q \in U_\alpha$,*

$$J_q\left(\frac{\partial}{\partial x^{2i-1}}\right)_q = \left(\frac{\partial}{\partial x^{2i}}\right)_q, \quad J_q\left(\frac{\partial}{\partial x^{2i}}\right)_q = -\left(\frac{\partial}{\partial x^{2i-1}}\right)_q$$

*are satisfied for $i = 1, \ldots, n$. Then $M$ is a complex manifold and $J$ is an almost complex structure which is induced from the complex structure of $M$.*

*Proof.* Let $(U_\alpha; x^1, \ldots, x^{2n})$, $(U_\beta; u^1, \ldots, u^{2n}) \in \mathcal{U}$ such that $U_\alpha \cap U_\beta \neq \emptyset$. Then we have

$$\frac{\partial}{\partial x^{2i-1}} = \sum_{j=1}^{n} \left( \frac{\partial u^{2j-1}}{\partial x^{2i-1}} \frac{\partial}{\partial u^{2j-1}} + \frac{\partial u^{2j}}{\partial x^{2i-1}} \frac{\partial}{\partial u^{2j}} \right),$$

$$\frac{\partial}{\partial x^{2i}} = \sum_{j=1}^{n} \left( \frac{\partial u^{2j-1}}{\partial x^{2i}} \frac{\partial}{\partial u^{2j-1}} + \frac{\partial u^{2j}}{\partial x^{2i}} \frac{\partial}{\partial u^{2j}} \right). \tag{2.7}$$

Applying $J$ to the above two equations, we have

$$\frac{\partial}{\partial x^{2i}} = \sum_{j=1}^{n} \left( \frac{\partial u^{2j-1}}{\partial x^{2i-1}} \frac{\partial}{\partial u^{2j}} - \frac{\partial u^{2j}}{\partial x^{2i-1}} \frac{\partial}{\partial u^{2j-1}} \right), \tag{2.8}$$

$$-\frac{\partial}{\partial x^{2i-1}} = \sum_{j=1}^{n} \left( \frac{\partial u^{2j-1}}{\partial x^{2i}} \frac{\partial}{\partial u^{2j}} - \frac{\partial u^{2j}}{\partial x^{2i}} \frac{\partial}{\partial u^{2j-1}} \right).$$

Comparing relations (2.8) and (2.9), we conclude

$$\frac{\partial u^{2j-1}}{\partial x^{2i-1}} = \frac{\partial u^{2j}}{\partial x^{2i}}, \quad \frac{\partial u^{2j-1}}{\partial x^{2i}} = -\frac{\partial u^{2j}}{\partial x^{2i-1}}. \tag{2.9}$$

We put

$$z^i = x^{2i-1} + \sqrt{-1}x^{2i},$$
$$w^i = u^{2i-1} + \sqrt{-1}u^{2i}.$$

Then $(z^1, \ldots, z^n)$ and $(w^1, \ldots, w^n)$ are complex coordinates in $U_\alpha$ and $U_\beta$, respectively, and in $U_\alpha \cap V_\beta$ it follows

$$w^k = f^k(z^1, \ldots, z^n), \quad f^k = \phi^k + \sqrt{-1}\psi^k,$$

where

$$\phi^k(x^1, \ldots, x^{2n}) = u^{2k-1}, \quad \psi^k(x^1, \ldots, x^{2n}) = u^{2k}.$$

Hence, from (2.9), we deduce that $f^k$ is holomorphic with respect to $z^i$ and therefore $M$ is a complex manifold.     $\square$

# 3

# Complex vector spaces, complexification

In this section we recall some algebraic results on complex vector spaces, applied to tangent and cotangent spaces of complex manifolds.

For the tangent space $T_x(M)$ at $x \in M$, we put

$$T_x^C(M) = \left\{ X_x + \sqrt{-1} Y_x \,|\, X_x, Y_x \in T_x(M) \right\}$$

and $T_x^C(M)$ is called the *complexification* of $T_x(M)$. We define

$$(X_x + \sqrt{-1} Y_x) + (X_x' + \sqrt{-1} Y_x') = (X_x + X_x') + \sqrt{-1}(Y_x + Y_x')$$

and for $\mathbf{C} \ni c = a + \sqrt{-1} b$,

$$c(X_x + \sqrt{-1} Y_x) = (aX_x - bY_x) + \sqrt{-1}(bX_x + aY_x).$$

Then $T_x^C(M)$ becomes a complex vector space. Identifying $T_x(M)$ with

$$\{ X_x + \sqrt{-1} 0_x \,|\, X_x \in T_x(M) \},$$

we regard that $T_x(M)$ is a subspace of $T_x^C(M)$. For $Z_x = X_x + \sqrt{-1} Y_x$, the complex conjugate $\overline{Z_x}$ of $Z_x$ is $\overline{Z_x} = X_x - \sqrt{-1} Y_x$. From this definition, we easily see that for $Z_x, W_x \in T_x^C(M)$,

$$\overline{Z_x + W_x} = \overline{Z_x} + \overline{W_x}, \quad \overline{c Z_x} = \bar{c} \overline{Z_x}.$$

For a linear transformation $A : T_x(M) \to T_x(M)$, we put

$$A(X_x + \sqrt{-1} Y_x) = AX_x + \sqrt{-1} AY_x.$$

Then $A$ defines a linear transformation on $T_x^C(M)$ and it satisfies

$$\overline{A Z_x} = A \overline{Z_x}, \quad \text{for} \quad Z_x \in T_x^C(M).$$

**Proposition 3.1.** $\dim_{\mathbf{C}} T_x^C(M) = \dim_{\mathbf{R}} M$.

M. Djorić, M. Okumura, *CR Submanifolds of Complex Projective Space*, Developments in Mathematics 19, DOI 10.1007/978-1-4419-0434-8_3, © Springer Science+Business Media, LLC 2010

*Proof.* Let $\{e_1, \ldots, e_n\}$ be a basis of $T_x(M)$ and $Z_x \in T_x^C(M)$. Then

$$Z_x = X_x + \sqrt{-1}Y_x = \sum_{i=1}^n X^i e_i + \sqrt{-1}\sum_{i=1}^n Y^i e_i = \sum_{i=1}^n (X^i + \sqrt{-1}Y^i)e_i.$$

Thus, $\{e_1, \ldots, e_n\}$ is a basis of $T_x^C(M)$ and $\dim_{\mathbf{C}} T_x^C(M) = n$. $\quad\square$

For a complex differentiable function $f = f_1 + \sqrt{-1}f_2$, we define the *derivative of $f$ by a complex tangent vector* $X_x + \sqrt{-1}Y_x$ by

$$(X_x + \sqrt{-1}Y_x)f = (X_x f_1 - Y_x f_2) + \sqrt{-1}(X_x f_2 + Y_x f_1)$$

and the bracket of complex vector fields by

$$[X + \sqrt{-1}Y, X' + \sqrt{-1}Y'] = [X, X'] - [Y, Y'] + \sqrt{-1}([X, Y'] + [Y, X']).$$

Then we have $[\overline{Z}, \overline{W}] = \overline{[Z, W]}$.

Let $(M, J)$ be an almost complex manifold with almost complex structure $J$. Then $J_x$ can be extended as an isomorphism of $T_x^C(M)$. We define $T_x^{(0,1)}(M)$ and $T_x^{(1,0)}(M)$ respectively by

$$T_x^{(0,1)}(M) = \{X_x + \sqrt{-1}J_x X_x | X_x \in T_x(M)\},$$

$$T_x^{(1,0)}(M) = \{X_x - \sqrt{-1}J_x X_x | X_x \in T_x(M)\}.$$

Then, we have

**Proposition 3.2.** *Under the above assumptions,*

$$T_x^C(M) = T_x^{(0,1)}(M) \oplus T_x^{(1,0)}(M),$$

*where $\oplus$ denotes the direct sum.*

*Proof.* For any $Z_x \in T_x^C(M)$, it follows

$$\begin{aligned}
Z_x &= \frac{1}{2}(Z_x + \sqrt{-1}J_x Z_x) + \frac{1}{2}(Z_x - \sqrt{-1}J_x Z_x) \\
&= \frac{1}{2}\left(X_x + \sqrt{-1}Y_x + \sqrt{-1}J_x(X_x + \sqrt{-1}Y_x)\right) \\
&\quad + \frac{1}{2}\left(X_x + \sqrt{-1}Y_x - \sqrt{-1}J_x(X_x + \sqrt{-1}Y_x)\right) \\
&= \frac{1}{2}\left((X_x - J_x Y_x) + \sqrt{-1}J_x(X_x - J_x Y_x)\right) \\
&\quad + \frac{1}{2}\left((X_x + J_x Y_x) - \sqrt{-1}J_x(X_x + J_x Y_x)\right),
\end{aligned}$$

where $\frac{1}{2}(X_x - J_x Y_x)$, $\frac{1}{2}(X_x + J_x Y_x) \in T_x(M)$.

If $Z_x \in T_x^{(0,1)}(M) \cap T_x^{(1,0)}(M)$, we have

$$Z_x = X_x + \sqrt{-1}J_xX_x = Y_x - \sqrt{-1}J_xY_x,$$

from which, $X_x = Y_x$ and $J_xX_x = -J_xY_x$. Applying $J_x$ to the last equation, we have $-X_x = Y_x$, which implies $X_x = Y_x = 0$ and $Z_x = 0$. Thus we have the direct sum. □

We note that $Z_x \in T_x^{(0,1)}(M)$ if and only if $J_xZ_x = -\sqrt{-1}Z_x$ and that $Z_x \in T_x^{(1,0)}(M)$ if and only if $J_xZ_x = \sqrt{-1}Z_x$.

**Definition 3.1.** A vector field $Z : M \ni x \longmapsto Z_x$ is said to be a vector field of type $(0,1)$ if $Z_x \in T_x^{(0,1)}(M)$ and of type $(1,0)$ if $Z_x \in T_x^{(1,0)}(M)$.

Let

$$T^C(M) = \bigcup_{x \in M} T_x^C(M), \quad T^{(0,1)}(M) = \bigcup_{x \in M} T_x^{(0,1)}(M),$$

$$T^{(1,0)}(M) = \bigcup_{x \in M} T_x^{(1,0)}(M).$$

$T^C(M)$ is a Lie algebra with respect to the bracket $[\ ,\ ]$ and

$$T^C(M) = T^{(0,1)}(M) \oplus T^{(1,0)}(M),$$

where $\oplus$ denotes the Whitney sum.

**Theorem 3.1.** $T^{(0,1)}(M)$ and $T^{(1,0)}(M)$ are involutive if and only if the Nijenhuis tensor $N$ vanishes identically.

*Proof.* First we note that for $Z \in T^{(0,1)}(M)$, $W \in T^{(1,0)}(M)$, it follows

$$JZ = -\sqrt{-1}Z, \quad JW = \sqrt{-1}W$$

and therefore

$$\begin{aligned}
N(Z,W) &= J[Z,W] - [JZ,W] - [Z,JW] - J[JZ,JW] \\
&= J[Z,W] + \sqrt{-1}[Z,W] - \sqrt{-1}[Z,W] - J[Z,W] = 0.
\end{aligned}$$

Let $Z,W \in T^{(0,1)}(M)$. Then

$$\begin{aligned}
N(Z,W) &= J[Z,W] - [-\sqrt{-1}Z,W] - [Z,-\sqrt{-1}W] - J[-\sqrt{-1}Z,-\sqrt{-1}W] \\
&= J[Z,W] + \sqrt{-1}[Z,W] + \sqrt{-1}[Z,W] + J[Z,W] \\
&= 2(J[Z,W] + \sqrt{-1}[Z,W]).
\end{aligned}$$

Thus

$$N(Z,W) = 0 \quad \text{if and only if} \quad J[Z,W] = -\sqrt{-1}[Z,W],$$

that is, $[Z,W] \in T^{(0,1)}(M)$. In a similar way we can prove the case of $T^{(1,0)}(M)$, which completes the proof. □

Let $M$ be an $n$-dimensional complex manifold and let $(z^1, \ldots, z^n)$ be complex coordinates in a neighborhood $U \ni x$. We regard that $M$ is a $2n$-dimensional differentiable manifold with local coordinates $(x^1, y^1, \ldots, x^n, y^n)$, where $z^i = x^i + \sqrt{-1}y^i$. Then $(\ldots, \frac{\partial}{\partial x^i}, \frac{\partial}{\partial y^i}, \ldots)$ is a basis of $T_x(M)$ and also a basis of $T_x^C(M)$. By definition of $J$ which is induced from the complex structure of $M$, it follows

$$\frac{\partial}{\partial x^i} = \frac{1}{2}\left(\frac{\partial}{\partial x^i} + \sqrt{-1}J\left(\frac{\partial}{\partial x^i}\right)\right) + \frac{1}{2}\left(\frac{\partial}{\partial x^i} - \sqrt{-1}J\left(\frac{\partial}{\partial x^i}\right)\right)$$
$$= \frac{1}{2}\left(\frac{\partial}{\partial x^i} + \sqrt{-1}\frac{\partial}{\partial y^i}\right) + \frac{1}{2}\left(\frac{\partial}{\partial x^i} - \sqrt{-1}\frac{\partial}{\partial y^i}\right).$$

We put

$$\frac{\partial}{\partial z^i} = \frac{1}{2}\left(\frac{\partial}{\partial x^i} + \sqrt{-1}\frac{\partial}{\partial y^i}\right), \qquad \frac{\partial}{\partial \bar{z}^i} = \frac{1}{2}\left(\frac{\partial}{\partial x^i} - \sqrt{-1}\frac{\partial}{\partial y^i}\right) \qquad (3.1)$$

which yields

$$\frac{\partial}{\partial x^i} = \frac{\partial}{\partial z^i} + \frac{\partial}{\partial \bar{z}^i}, \qquad \frac{\partial}{\partial y^i} = \sqrt{-1}\left(\frac{\partial}{\partial z^i} - \frac{\partial}{\partial \bar{z}^i}\right), \qquad \overline{\frac{\partial}{\partial z^i}} = \frac{\partial}{\partial \bar{z}^i}.$$

From (3.1), we know that any $X \in T_x^C(M)$ can be expressed as a linear combination of $\frac{\partial}{\partial z^i}$ and $\frac{\partial}{\partial \bar{z}^i}$, $i = 1, \ldots, n$. On the other hand, suppose that

$$\sum_{i=1}^n \left(a^i \frac{\partial}{\partial z^i} + b^i \frac{\partial}{\partial \bar{z}^i}\right) = 0.$$

Then, from (3.1), it follows

$$\sum_{i=1}^n (a^i + b^i)\frac{\partial}{\partial x^i} + \sqrt{-1}\sum_{i=1}^n (a^i - b^i)\frac{\partial}{\partial y^i} = 0.$$

Hence, $a^i + b^i = 0$ and $a^i - b^i = 0$ which implies $a^i = b^i = 0$ for $i = 1, \ldots, n$. Therefore we conclude that

$$\left\{\left(\frac{\partial}{\partial z^1}\right)_x, \ldots, \left(\frac{\partial}{\partial z^n}\right)_x, \left(\frac{\partial}{\partial \bar{z}^1}\right)_x, \ldots, \left(\frac{\partial}{\partial \bar{z}^n}\right)_x\right\}$$

forms a basis of $T_x^C(M)$.

For a natural basis of a tangent space $T_x(M)$ at $x \in M$ we consider its dual basis $\{(dx^1)_x, (dy^1)_x, \ldots, (dx^n)_x, (dy^n)_x\}$ in the cotangent space $T_x(M)^*$ and we put

$$(dz^i)_x = (dx^i)_x + \sqrt{-1}(dy^i)_x, \qquad (d\bar{z}^i)_x = (dx^i)_x - \sqrt{-1}(dy^i)_x. \qquad (3.2)$$

Consequently, it follows

$$(dz^i)_x \left( \frac{\partial}{\partial z^j} \right)_x = \frac{1}{2}(dx^i + \sqrt{-1}dy^i)_x \left( \frac{\partial}{\partial x^j} - \sqrt{-1}\frac{\partial}{\partial y^j} \right)_x = \frac{1}{2}(\delta_j^i + \delta_j^i) = \delta_j^i,$$

$$(dz^i)_x \left( \frac{\partial}{\partial \bar{z}^j} \right)_x = \frac{1}{2}(dx^i + \sqrt{-1}dy^i)_x \left( \frac{\partial}{\partial x^j} + \sqrt{-1}\frac{\partial}{\partial y^j} \right)_x = 0.$$

In the same way we have

$$(d\bar{z}^i)_x \left( \frac{\partial}{\partial z^j} \right)_x = 0, \qquad (d\bar{z}^i)_x \left( \frac{\partial}{\partial \bar{z}^j} \right)_x = \delta_j^i.$$

This shows that

$$\left\{ (dz^1)_x, (d\bar{z}^1)_x, \ldots, (dz^n)_x, (d\bar{z}^n)_x \right\} \tag{3.3}$$

is the dual basis of

$$\left\{ \left( \frac{\partial}{\partial z^1} \right)_x, \left( \frac{\partial}{\partial \bar{z}^1} \right)_x, \ldots, \left( \frac{\partial}{\partial z^n} \right)_x, \left( \frac{\partial}{\partial \bar{z}^n} \right)_x \right\}. \tag{3.4}$$

*Remark* 3.1. Note that the reason why the minus sign and the factor $\frac{1}{2}$ appear in (3.1) is because we wanted to choose the basis (3.3) to be dual to (3.4) defined by (3.2).

We write for a $C^\infty$ function $f$ defined on a neighborhood of $x \in M$, $(\frac{\partial}{\partial z^i})f = \frac{\partial f}{\partial z^i}$ and $(\frac{\partial}{\partial \bar{z}^i})f = \frac{\partial f}{\partial \bar{z}^i}$. We have

$$df \left( \frac{\partial}{\partial z^i} \right) = \frac{\partial f}{\partial z^i}, \quad df \left( \frac{\partial}{\partial \bar{z}^i} \right) = \frac{\partial f}{\partial \bar{z}^i}$$

and therefore

$$df = \sum_{i=1}^n \left( \frac{\partial f}{\partial z^i} dz^i + \frac{\partial f}{\partial \bar{z}^i} d\bar{z}^i \right). \tag{3.5}$$

**Definition 3.2.** Let $r$ be a positive integer such that $r = p + q$ where $p$, $q$ are nonnegative integers. Let an $r$-form $\omega$ on $M$ be spanned by the set $\{dz^{i_1} \wedge \cdots \wedge dz^{i_p} \wedge d\bar{z}^{j_1} \wedge \cdots \wedge d\bar{z}^{j_q}\}$, where $\{i_1, \ldots, i_p\}$ and $\{j_1, \ldots, j_q\}$ run over the set of all increasing multi-indices of length $p$ and $q$. Then $\omega$ is called a *complex differential form* of type $(p, q)$.

Since an $r$-form $\omega$ of type $(p, q)$, we have just defined, can be expressed as

$$\omega = \sum_{\substack{i_1 < \cdots < i_p \\ j_1 < \cdots < j_q}} \omega_{i_1 \ldots i_p j_1 \ldots j_q} dz^{i_1} \wedge \cdots \wedge dz^{i_p} \wedge d\bar{z}^{j_1} \wedge \cdots \wedge d\bar{z}^{j_q}, \tag{3.6}$$

using (3.2), we can easily prove the following

**Proposition 3.3.** *Let $\omega$ and $\eta$ be complex differential forms.*

(1)    *If $\omega$ is of type $(p, q)$, then $\bar\omega$ is of type $(q, p)$.*

(2)    *If $\omega$ is of type $(p, q)$ and $\eta$ is of type $(p', q')$, then $\omega \wedge \eta$ is of type $(p + p', q + q')$.*

Further, using (3.5) and (3.6), we compute the exterior differential $d\omega$ of any complex $r$-form $\omega$ of type $(p, q)$.

$$
d\omega = \sum_{\substack{i_1 < \cdots < i_p \\ j_1 < \cdots < j_q}} \sum_{k=1}^{n} \left( \frac{\partial \omega_{i_1 \ldots i_p j_1 \ldots j_q}}{\partial z^k} dz^k + \frac{\partial \omega_{i_1 \ldots i_p j_1 \ldots j_q}}{\partial \bar{z}^k} d\bar{z}^k \right) \wedge dz^{i_1} \wedge \ldots
$$

$$
\cdots \wedge dz^{i_p} \wedge d\bar{z}^{j_1} \wedge \cdots \wedge d\bar{z}^{j_q}
$$

$$
= \sum_{\substack{i_1 < \cdots < i_{p+1} \\ j_1 < \cdots < j_q}} \sum_{s=1}^{p+1} (-1)^{s-1} \frac{\partial \omega_{i_1 \ldots \hat{i}_s \ldots i_{p+1} j_1 \ldots j_q}}{\partial z^{i_s}} dz^{i_1} \wedge \ldots
$$

$$
\cdots \wedge dz^{i_{p+1}} \wedge d\bar{z}^{j_1} \wedge \cdots \wedge d\bar{z}^{j_q}
$$

$$
+ (-1)^p \sum_{\substack{i_1 < \cdots < i_p \\ j_1 < \cdots < j_{q+1}}} \sum_{t=1}^{q+1} (-1)^{t-1} \frac{\partial \omega_{i_1 \ldots i_p j_1 \ldots \hat{j}_t \ldots j_{q+1}}}{\partial \bar{z}^{j_t}} dz^{i_1} \wedge \ldots
$$

$$
\cdots \wedge dz^{i_p} \wedge d\bar{z}^{j_1} \wedge \cdots \wedge d\bar{z}^{j_{q+1}}.
$$

Therefore, $d\omega$ is expressed as a sum of $(r + 1)$-forms of type $(p + 1, q)$ and of type $(p, q + 1)$, denoted respectively by $\partial\omega$ and $\bar\partial\omega$. Thus we obtain two differential operators $\partial$ and $\bar\partial$ and this information is written as

$$
d\omega = \partial\omega + \bar\partial\omega, \quad d = \partial + \bar\partial. \tag{3.7}
$$

**Proposition 3.4.** *Let $\omega$, $\eta$ be $r$-forms on $M$ and $a \in \mathbf{C}$. Then we have*

$$
\partial(\omega + \eta) = \partial\omega + \partial\eta, \quad \bar\partial(\omega + \eta) = \bar\partial\omega + \bar\partial\eta, \quad \partial(a\omega) = a\partial\omega, \quad \bar\partial(a\omega) = a\bar\partial\omega.
$$

*Proof.* It is sufficient to prove the above relations for an $r$-form of type $(p, q)$. We compare the following two equations:

$$
d(\omega + \eta) = d\omega + d\eta = \partial\omega + \partial\eta + \bar\partial\omega + \bar\partial\eta,
$$
$$
d(\omega + \eta) = \partial(\omega + \eta) + \bar\partial(\omega + \eta).
$$

Since $\partial\omega + \partial\eta$ is of type $(p + 1, q)$ and $\bar\partial\omega + \bar\partial\eta$ is of type $(p, q + 1)$, the first two relations are satisfied. Similarly, we can prove the other relations.    □

**Proposition 3.5.** *For differential operators $\partial$, $\bar\partial$ and $r$-form $\omega$, we have*

$$
\partial^2\omega = 0, \quad (\partial\bar\partial + \bar\partial\partial)\omega = 0, \quad \bar\partial^2\omega = 0, \quad \overline{\partial\omega} = \bar\partial\bar\omega, \quad \overline{\bar\partial\omega} = \partial\bar\omega.
$$

*Proof.* Since $d^2 = 0$, using (3.7) and Proposition 3.4, we compute

$$0 = d^2\omega = d(\partial\omega + \bar{\partial}\omega)$$
$$= \partial(\partial\omega + \bar{\partial}\omega) + \bar{\partial}(\partial\omega + \bar{\partial}\omega)$$
$$= \partial^2\omega + (\partial\bar{\partial} + \bar{\partial}\partial)\omega + \bar{\partial}^2\omega.$$

As $\partial^2\omega$ is of type $(p+2,q)$, $(\partial\bar{\partial} + \bar{\partial}\partial)\omega$ is of type $(p+1,q+1)$ and $\bar{\partial}^2\omega$ is of type $(p,q+2)$, we conclude that each of them vanishes.

To prove the other relations, we remember the definition of $d\bar{\omega}$, that is, $d\bar{\omega} = \overline{d\omega}$. Therefore, $d\bar{\omega} = \overline{\partial\omega + \bar{\partial}\omega}$. On the other hand, using (3.7), it follows $d\bar{\omega} = \partial\bar{\omega} + \bar{\partial}\bar{\omega}$. Comparing the type of the right hand members of the last two equations, we get the other two relations of the proposition.       □

**Theorem 3.2.** *Let $f$ be a function defined on an open set of $M$. Then the following three conditions are equivalent:*

(1) $\bar{\partial}f = 0$ ;      (2) $\frac{\partial f}{\partial \bar{z}^i} = 0$ *for* $i = 1,\ldots,n$ ;      (3) $f$ *is holomorphic.*

*Proof.* Since $\bar{\partial}f = \sum_{i=1}^n \frac{\partial f}{\partial \bar{z}^i} d\bar{z}^i$, condition (1) is equivalent to (2).

We put $f = f_1 + \sqrt{-1}f_2$ and $z^i = x^i + \sqrt{-1}y^i$. Then $\frac{\partial f}{\partial \bar{z}^i} = 0$ is equivalent to $\frac{\partial f}{\partial x^i} = -\sqrt{-1}\frac{\partial f}{\partial y^i}$. Hence,

$$\frac{\partial}{\partial x^i}(f_1 + \sqrt{-1}f_2) = -\sqrt{-1}\frac{\partial}{\partial y^i}(f_1 + \sqrt{-1}f_2).$$

This implies that $\frac{\partial f_1}{\partial x^i} = \frac{\partial f_2}{\partial y^i}$ and $\frac{\partial f_2}{\partial x^i} = -\frac{\partial f_1}{\partial y^i}$ for $i = 1,\ldots,n$. These are the Cauchy-Riemann equations and therefore $f$ is holomorphic.       □

**Definition 3.3.** A $p$-form $\omega$ is said to be a *holomorphic $p$-form* if $\omega$ is of type $(p,0)$ and $d\omega$ is of type $(p+1,0)$.

This is equivalent to saying that $\omega$ is of type $(p,0)$ and $\bar{\partial}\omega = 0$.

# 4

# Kähler manifolds

Kähler manifolds are the most studied among complex manifolds. In this section we provide basic material about these manifolds and we present several examples. In particular, we prove that the complex projective space is a Kähler manifold.

**Definition 4.1.** Let $(M, J)$ be an almost complex manifold. If a Riemannian metric $g$ satisfies

$$g(X, Y) = g(JX, JY) \tag{4.1}$$

for any $X$, $Y \in T(M)$, $g$ is said to be a *Hermitian metric* and the almost complex manifold $(M, J)$ with Hermitian metric $g$ is said to be an *almost Hermitian manifold*.

Therefore, using (4.1), we conclude

$$g(JX, Y) = g(J^2 X, JY) = -g(X, JY),$$

which means that $J$ is skew-symmetric.

Moreover, we prove

**Theorem 4.1.** *On any almost complex manifold, there exists a Hermitian metric.*

*Proof.* On $M$ there exists a Riemannian metric $g'$. We put

$$g(X, Y) = \frac{1}{2}\{g'(X, Y) + g'(JX, JY)\}.$$

It is easy to see that $g$ is a Hermitian metric. □

Let $(M, J)$ be an almost Hermitian manifold with Hermitian metric $g$. The *fundamental 2-form, Kähler form* $\Omega$ of $M$ is defined by

$$\Omega(X, Y) = g(JX, Y) \tag{4.2}$$

for all vector fields $X$ and $Y$ on $M$.

M. Djorić, M. Okumura, *CR Submanifolds of Complex Projective Space*,
Developments in Mathematics 19, DOI 10.1007/978-1-4419-0434-8_4,
© Springer Science+Business Media, LLC 2010

**Lemma 4.1.** $\Omega$ *is skew symmetric, that is,* $\Omega(X,Y) = -\Omega(Y,X)$.

*Proof.* Using (4.1) and (4.2) we compute

$$\Omega(Y,X) = g(JY,X) = g(J^2Y, JX) = -g(Y, JX) = -\Omega(X,Y). \qquad \square$$

Let $\nabla$ be the *Levi-Civita connection* with respect to the Hermitian metric $g$, that is, $\nabla$ satisfies $\nabla g = 0$ and $[X,Y] = \nabla_X Y - \nabla_Y X$. We now express the Nijenhuis tensor $N$ using the Levi-Civita connection $\nabla$ with respect to the Hermitian metric $g$.

$$
\begin{aligned}
N(X,Y) &= J[X,Y] - [JX,Y] - [X,JY] - J[JX,JY] \\
&= J(\nabla_X Y - \nabla_Y X) - \nabla_{JX} Y + \nabla_Y (JX) \\
&\quad - \nabla_X(JY) + \nabla_{JY} X - J(\nabla_{JX}(JY) - \nabla_{JY}(JX)) \\
&= J\nabla_X Y - J\nabla_Y X - \nabla_{JX} Y + (\nabla_Y J)X + J\nabla_Y X - (\nabla_X J)Y \\
&\quad - J\nabla_X Y + \nabla_{JY} X - J(\nabla_{JX} J)Y + \nabla_{JX} Y + J(\nabla_{JY} J)X - \nabla_{JY} X.
\end{aligned}
$$

Thus we have

$$N(X,Y) = (\nabla_Y J)X - (\nabla_X J)Y + J(\nabla_{JY} J)X - J(\nabla_{JX} J)Y. \qquad (4.3)$$

**Definition 4.2.** If a complex manifold $(M,J)$ with Hermitian metric $g$ satisfies $d\Omega = 0$, then $(M,J)$ is called a *Kähler manifold* and the metric $g$ is called a *Kähler metric*.

**Theorem 4.2.** *A necessary and sufficient condition that a complex manifold* $(M,J)$ *with Hermitian metric is a Kähler manifold is* $\nabla_X J = 0$ *for any* $X \in T(M)$.

*Proof.* Since for a $p$-form $\omega$ we have

$$d\omega(X_1,\dots,X_{p+1}) = \sum_{k=1}^{p+1} (-1)^{k-1}(\nabla_{X_k}\omega)(X_1,\dots,\hat{X}_k,\dots,X_{p+1}) \qquad (4.4)$$

it follows

$$d\Omega(X,Y,Z) = (\nabla_X \Omega)(Y,Z) - (\nabla_Y \Omega)(X,Z) + (\nabla_Z \Omega)(X,Y).$$

On the other hand,

$$
\begin{aligned}
(\nabla_X \Omega)(Y,Z) &= \nabla_X(\Omega(Y,Z)) - \Omega(\nabla_X Y, Z) - \Omega(Y, \nabla_X Z) \\
&= \nabla_X(g(JY,Z)) - g(J\nabla_X Y, Z) - g(JY, \nabla_X Z) \\
&= g((\nabla_X J)Y, Z).
\end{aligned}
$$

Hence we have

$$d\Omega(X,Y,Z) = g((\nabla_X J)Y, Z) - g((\nabla_Y J)X, Z) + g((\nabla_Z J)X, Y).$$

Thus the sufficiency is obvious. To prove the necessity, we note that $J(JX) = J^2 X = -X$. Differentiating covariantly this equation, we have

$$(\nabla_Y J)JX + J(\nabla_Y J)X + J^2 \nabla_Y X = -\nabla_Y X,$$

from which it follows

$$(\nabla_Y J)JX = -J(\nabla_Y J)X. \qquad (4.5)$$

Making use of (4.5), we compute

$$
\begin{aligned}
d\Omega(JX,&Y, JZ) - d\Omega(JY, X, JZ) \\
&= g((\nabla_{JX} J)Y, JZ) - g((\nabla_Y J)JX, JZ) + g((\nabla_{JZ} J)JX, Y) \\
&\quad - g((\nabla_{JY} J)X, JZ) + g((\nabla_X J)JY, JZ) - g((\nabla_{JZ} J)JY, X) \\
&= g(J(\nabla_{JY} J)X - J(\nabla_{JX} J)Y, Z) + g(J(\nabla_Y J)X, JZ) \\
&\quad - g(J(\nabla_X J)Y, JZ) + g((\nabla_{JZ} J)JX, Y) - g((\nabla_{JZ} J)JY, X) \\
&= g(N(X,Y), Z) - g((\nabla_Y J)X - (\nabla_X J)Y, Z) \\
&\quad + g(J(\nabla_Y J)X, JZ) - g(J(\nabla_X J)Y, JZ) \\
&\quad + g((\nabla_{JZ} J)JX, Y) - g((\nabla_{JZ} J)JY, X) \\
&= g(N(X,Y), Z) - g(J(\nabla_{JZ} J)X, Y) + g(J(\nabla_{JZ} J)Y, X) \\
&= g(N(X,Y), Z) + 2g(J(\nabla_{JZ} J)Y, X).
\end{aligned}
$$

Thus, $N = 0$ and $d\Omega = 0$ imply that $J(\nabla_{JZ} J) = 0$. Since $J$ is isomorphism, this implies that $\nabla_X J = 0$ for any $X \in T(M)$, which completes the proof. $\square$

Now we give some examples of Kähler manifolds.

*Example 4.1.* Any complex manifold $M$ of $\dim_{\mathbf{C}} M = 1$ is Kähler manifold.

The Kähler form $\Omega$ is a 2-form and therefore $d\Omega$ is a 3-form. But $\dim_{\mathbf{R}} M = 2\dim_{\mathbf{C}} M = 2$. Hence $d\Omega$ vanishes identically. $\diamond$

*Example 4.2. n-dimensional complex space $\mathbf{C}^n$.*

Since $\mathbf{C}^n$ can be identified with $\mathbf{R}^{2n}$, let $\langle,\rangle$ be the Euclidean metric of $\mathbf{R}^{2n}$. Then we have

$$\left\langle \frac{\partial}{\partial x^i}, \frac{\partial}{\partial x^j} \right\rangle = \left\langle \frac{\partial}{\partial y^i}, \frac{\partial}{\partial y^j} \right\rangle = \delta_{ij}, \qquad \left\langle \frac{\partial}{\partial x^i}, \frac{\partial}{\partial y^j} \right\rangle = 0.$$

This, together with $J\left(\frac{\partial}{\partial x^i}\right) = \frac{\partial}{\partial y^i}$ and $J\left(\frac{\partial}{\partial y^i}\right) = -\frac{\partial}{\partial x^i}$, implies that $\langle,\rangle$ is a Hermitian metric of $(\mathbf{C}^n, J)$. We put

$$\Omega = \sum_{k,l=1}^{n} (a_{kl} dx^k \wedge dx^l + b_{kl} dx^k \wedge dy^l + c_{kl} dy^k \wedge dy^l)$$

and note that

$$dx^k \wedge dy^l \left( \frac{\partial}{\partial x^i}, \frac{\partial}{\partial x^j} \right) = dx^k \wedge dy^l \left( \frac{\partial}{\partial y^i}, \frac{\partial}{\partial y^j} \right) = 0,$$

$$dx^k \wedge dy^l \left( \frac{\partial}{\partial x^i}, \frac{\partial}{\partial y^j} \right) = dx^k \left( \frac{\partial}{\partial x^i} \right) dy^l \left( \frac{\partial}{\partial y^j} \right) - dx^k \left( \frac{\partial}{\partial y^j} \right) dy^l \left( \frac{\partial}{\partial x^i} \right)$$

$$= \delta_i^k \delta_j^l.$$

Then

$$\Omega \left( \frac{\partial}{\partial x^i}, \frac{\partial}{\partial x^j} \right) = \sum_{k,l}^{n} a_{kl} \delta_i^k \delta_j^l = a_{ij}.$$

On the other hand, it follows

$$\Omega \left( \frac{\partial}{\partial x^i}, \frac{\partial}{\partial x^j} \right) = \left\langle J \left( \frac{\partial}{\partial x^i} \right), \frac{\partial}{\partial x^j} \right\rangle = \left\langle \frac{\partial}{\partial y^i}, \frac{\partial}{\partial x^j} \right\rangle = 0.$$

Hence we have $a_{ij} = 0$. In entirely the same way, we conclude $b_{ij} = \delta_{ij}$ and $c_{ij} = 0$. Thus, the Kähler form $\Omega$ of $(\mathbf{C}^n, J)$ is represented by

$$\Omega = \sum_{k=1}^{n} dx^k \wedge dy^k. \tag{4.6}$$

From (4.6) we conclude that $d\Omega = 0$ and that $(\mathbf{C}^n, J)$, with usual Euclidean metric, is a Kähler manifold.                                   $\diamond$

*Example 4.3. Complex projective space* $\mathbf{P}^n(\mathbf{C})$.

According to notation from Example 1.3, we express $\mathbf{P}^n(\mathbf{C})$ by

$$\{[(z^0, z^1, \ldots, z^n)] \mid z^i \in \mathbf{C}, \ i = 0, \ldots, n\}.$$

Let $U_j$ be an open subset of $\mathbf{P}^n(\mathbf{C})$ defined by $z^j \neq 0$ and put $t_j^k = \frac{z^k}{z^j}$, $j, k = 0, 1, \ldots, n$ on $U_j$. Then

$$(t_j^0, t_j^1, \ldots, t_j^{j-1}, t_j^{j+1}, \ldots, t_j^n)$$

is a local coordinate system in $U_j$ and we put

$$f_j = \sum_{k=0}^{n} t_j^k \bar{t}_j^k = \sum_{k=0}^{n} |t_j^k|^2.$$

Then, on $U_j \cap U_k$, it follows $f_j = f_k t_j^k \bar{t}_j^k$. Since $t_j^k$ is holomorphic on $U_j$, and in particular on $U_j \cap U_k$, it follows $\bar{\partial} \log t_j^k = 0$ and therefore

$$\partial\overline{\partial}\log \overline{t}_j^k = -\overline{\partial}\partial\log \overline{t}_j^k = -\overline{\partial}\,\overline{\partial}\log t_j^k = 0 \quad \text{on} \quad U_j \cap U_k.$$

Hence

$$\partial\overline{\partial}\log f_j = \partial\overline{\partial}\left(\log f_k + \log t_j^k + \log \overline{t}_j^k\right) = \partial\overline{\partial}\log f_k$$

on $U_j \cap U_k$, which shows that $\partial\overline{\partial}\log f_j$ does not depend on the choice of local coordinates. Moreover, $\partial\overline{\partial}\log f_j$ is a closed 2-form, since

$$d(\partial\overline{\partial}\log f_j) = (\partial + \overline{\partial})(\partial\overline{\partial}\log f_j) = 0.$$

Further we note that $\partial\overline{\partial}\log f_j$ is purely imaginary. In fact, $f_j$ is a real function and thus

$$\overline{f}_j = f_j, \ \log f_j = \overline{\log f_j}.$$

Therefore it follows

$$\overline{\partial(\overline{\partial}\log f_j)} = \overline{\partial}\,\overline{\overline{\partial}\log f_j} = \overline{\partial}\partial\overline{\log f_j} = -\partial\overline{\partial}\overline{\log f_j} = -\partial\overline{\partial}\log f_j$$

and consequently

$$\Omega = \sqrt{-1}\partial\overline{\partial}\log f_j \tag{4.7}$$

is a globally defined closed 2-form on $\mathbf{P}^n(\mathbf{C})$. Since

$$\frac{\partial^2 \log f_j}{\partial t_j^h \partial \overline{t}_j^i} = \frac{\partial^2 (\log \sum_k t_j^k \overline{t}_j^k)}{\partial t_j^k \partial \overline{t}_j^i} = \frac{(\sum_k t_j^k \overline{t}_j^k)\delta_{hi} - t_j^i \overline{t}_j^h}{(\sum_k t_j^k \overline{t}_j^k)^2},$$

the explicit expression of $\Omega$ is

$$\Omega = \sqrt{-1}\partial\overline{\partial}\log f_j = \sqrt{-1}\sum_{h,i=1}^{n} \frac{\partial^2 \log f_j}{\partial t_j^h \partial \overline{t}_j^i} dt_j^h \wedge d\overline{t}_j^i$$

$$= \sqrt{-1}\sum_{h,i=1}^{n} \frac{(\sum_k t_j^k \overline{t}_j^k)\delta_{hi} - t_j^i \overline{t}_j^h}{(\sum_k t_j^k \overline{t}_j^k)^2} dt_j^h \wedge d\overline{t}_j^i. \tag{4.8}$$

We put

$$g(X, Y) = \Omega(X, JY)$$

and show that $g$ is a Hermitian metric. First we show that $g$ is symmetric.

Let $X$ and $Y$ be real tangent vector fields of $\mathbf{P}^n(\mathbf{C})$. Then they can be expressed on $U_j$ respectively as

$$X = \sum_i \left(\alpha^i \frac{\partial}{\partial t_j^i} + \overline{\alpha}^i \frac{\partial}{\partial \overline{t}_j^i}\right), \quad Y = \sum_h \left(\beta^h \frac{\partial}{\partial t_j^h} + \overline{\beta}^h \frac{\partial}{\partial \overline{t}_j^h}\right).$$

Since $J(\frac{\partial}{\partial t_j^i}) = \sqrt{-1}\frac{\partial}{\partial t_j^i}$ and $J(\frac{\partial}{\partial \overline{t}_j^i}) = -\sqrt{-1}\frac{\partial}{\partial \overline{t}_j^i}$, we have

$$JX = \sqrt{-1} \sum_k \left( \alpha^k \frac{\partial}{\partial t_j^k} - \overline{\alpha}^k \frac{\partial}{\partial \overline{t}_j^k} \right).$$

Substituting this in (4.8), we obtain

$$g(X,Y) = \sqrt{-1} \sum_{h,i=1}^n \frac{\partial^2 \log f_j}{\partial t_j^h \partial \overline{t}_J^i} dt_j^h \wedge d\overline{t}_j^i (X, JY)$$

$$= \sum_{h,i=1}^n \frac{\partial^2 \log f_j}{\partial t_j^h \partial \overline{t}_j^i} (\alpha^h \overline{\beta}^i + \beta^h \overline{\alpha}^i) = g(Y, X)$$

and we conclude that $g$ is symmetric.

Further, from (4.8) and using the Schwarz inequality, it follows that $g$ is positive definite. Namely,

$$g(X,X) = 2 \frac{(\sum_k t_j^k \overline{t}_j^k)(\sum_i \alpha^i \overline{\alpha}^i) - \sum_{h,i}(t_j^i \overline{\alpha}^i)(\overline{t}_j^h \alpha^h)}{(\sum_k t_j^k \overline{t}_j^k)^2}$$

$$= 2 \frac{\sum_i |\alpha^i|^2 \sum_k |t_j^k|^2 - |\sum_i t_j^i \overline{\alpha}^i|^2}{(\sum_k |t_j^k|^2)^2} \geq 0,$$

We call this metric $g$ the *Fubini-Study metric*. Since

$$g(JX, JY) = \Omega(JX, J^2Y) = \Omega(JX, -Y) = \Omega(X, JY) = g(X, Y),$$

we conclude that $g$ is a Hermitian metric on $\mathbf{P}^n(\mathbf{C})$ and therefore $(\mathbf{P}^n(\mathbf{C}), J)$ is a Kähler manifold with a Kähler metric $g$.    $\diamond$

# 5

# Structure equations of a submanifold

A differentiable mapping $\imath$ of $M$ into $M'$ is called an *immersion* if $(\imath_*)_x$ is injective for every point $x$ of $M$. Here $\imath_*$ is the usual differential map $\imath_* : T_x(M) \rightarrow T_{\imath(x)}(M')$. We say then that $M$ is immersed in $M'$ by $\imath$ or that $M$ is an *immersed submanifold* of $M'$. When an immersion $\imath$ is injective, it is called an *embedding* of $M$ into $M'$. We say then that $M$ (or the image $\imath(M)$) is an *embedded submanifold* (or, simply, a *submanifold*) of $M'$. In this sense, throughout what follows, we adopt the convention that by submanifold we mean embedded submanifold. If the dimensions of $M$ and $M'$ are $n$ and $n + p$, respectively, the number $p$ is called the *codimension* of a submanifold $M$. The interested reader is referred to [5] and [33] for further information and more details.

In this section the reader will be reminded of some important properties of submanifolds, and some auxiliary results will be quoted or derived, such as the well-known Gauss and Weingarten formulae, the equation of Gauss, Codazzi and Ricci-Kühne.

Let $M$ and $M'$ be differentiable manifolds and $f$ be a differentiable map $f : M \rightarrow M'$. Note that for a given vector field $X$ on $M$, it follows that $f_* X$ is not always a vector field on $M'$. For this purpose, we first define the notion of a vector field along the map $f$.

**Definition 5.1.** A *vector field along the map* $f : M \rightarrow M'$ is a differentiable map

$$X' : M \ni x \mapsto X'_x \in T_{f(x)}(M') \quad \text{which} \quad \text{satisfies} \quad \pi(X'_x) = f(x),$$

for any $x \in M$, where $\pi$ is the natural projection $\pi : T(M') \rightarrow M'$.

*Example* 5.1. Let $x$ be a curve on $M'$ defined on an open interval $(a, b) = M$ and let $f$ be a differentiable map $f : M \rightarrow M'$, defined by

$$f : M \ni t \mapsto x(t) \in M'.$$

Then the tangent vector field $x'(t)$ of the curve $x$ is a vector field along $f$. $\Diamond$

M. Djorić, M. Okumura, *CR Submanifolds of Complex Projective Space*, Developments in Mathematics 19, DOI 10.1007/978-1-4419-0434-8_5, © Springer Science+Business Media, LLC 2010

*Example* 5.2. Let $X'$ be a given vector field on $M'$. For a differentiable map $f : M \to M'$, we can define a vector field on $M$, by $X_x = X'_{f(x)}$. Then $X'$ is a vector field along $f$.    ◇

*Example* 5.3. Let $X$ be a vector field on $M$. The vector field $X'$, which is obtained by $X'_x = (f_*)_x(X_x)$, $x \in M$, is a vector field along $f$ and we denote this vector field by $f_*X$. Example 5.1 is a special case of this example, that is, $x'_t = f_*(\frac{d}{dt})$.    ◇

We denote by $\mathfrak{X}(M)$ and $\mathfrak{X}_f$, the set of differentiable vector fields on $M$ and the set of vector fields along $f$, respectively.

The covariant differentiation $\nabla'_{X'}Y'$ with respect to given linear connection $\nabla'$ of $M'$ is defined for vector fields $X'$, $Y'$ on $M'$. Since the vector field along $f$ is not always the vector field on $M'$, we define the covariant differentiation along the map $f$.

**Definition 5.2.** Let $M'$ be a differentiable manifold with linear connection and $\nabla'$ be the covariant differentiation with respect to this connection. Then, a map

$$\mathfrak{X}(M) \times \mathfrak{X}_f \ni (Y, X) \mapsto \nabla'_Y X \in \mathfrak{X}_f$$

which satisfies the following properties (1)–(4) is determined uniquely and the map $\mathfrak{X}(M') \times \mathfrak{X}_f \to \mathfrak{X}_f$ is called a *covariant differentiation along* $f$.

(1) If $Y_1, Y_2 \in \mathfrak{X}(M)$, then

$$\nabla'_{Y_1+Y_2} X = \nabla'_{Y_1} X + \nabla'_{Y_2} X.$$

(2) For a function $\lambda$ on $M$,

$$\nabla'_{\lambda Y} X = \lambda \nabla'_Y X.$$

(3) For $X_1, X_2 \in \mathfrak{X}_f$ ,

$$\nabla'_Y (X_1 + X_2) = \nabla'_Y X_1 + \nabla'_Y X_2.$$

(4) For a function $\lambda'$ on $M'$,

$$\nabla'_Y (\lambda' X) = (Y\lambda')X + \lambda' \nabla'_Y X,$$

where $\lambda'X \in \mathfrak{X}_f$, since $(\lambda'X)_x = \lambda'(x)X_x$ for $x \in f(M) \subset M'$.

Now, let $M$ be an $n$-dimensional submanifold of an $(n + p)$-dimensional Riemannian manifold $(\overline{M}, \overline{g})$ and $\imath : M \to (\overline{M}, \overline{g})$ be the immersion. One more piece of notation: throughout the manuscript we also denote by $\imath$ the differential $\imath_*$ of the immersion, or we omit to mention $\imath$, for brevity of notation. Then, for $Y \in T(M)$, $\imath Y$ is a vector field along the immersion $\imath$.

Further, we define a Riemannian metric $g$ on $M$ by

$$g(X,Y) = \bar{g}(\imath X, \imath Y),$$

where $X, Y \in T(M)$. The Riemannian metric $g$ is called the *induced metric* from $\bar{g}$ and the immersion $\imath$ is called an *isometric immersion*. The tangent bundle $T(\overline{M})$ splits into the tangential part and the normal part to $M$, that is,

$$T(\overline{M}) = \imath T(M) \oplus T^\perp(M),$$

where $T(M) = \bigcup_{x \in M} T_x(M)$ is the tangent bundle of $M$ in $\overline{M}$, $T^\perp(M) = \bigcup_{x \in M} T^\perp_{\imath(x)}(M)$ is the *normal bundle* of $M$ in $\overline{M}$ and $T^\perp_{\imath(x)}(M)$ denotes the orthogonal complement of $\imath T_x(M)$ in $T_{\imath(x)}(\overline{M})$.

Let $\overline{\nabla}$ be the Levi-Civita connection of $(\overline{M}, \bar{g})$. Using Definition 5.2, regarding $\overline{\nabla}$ as a covariant differentiation along $\imath$, we can derive the following *Gauss formula*

$$\overline{\nabla}_X \imath Y = \imath \nabla_X Y + h(X,Y), \tag{5.1}$$

where $X, Y \in T(M)$. It is easily verified that $\nabla$ defines a connection of $M$ which is called the *induced connection* from $\overline{\nabla}$, while the normal part $h(X,Y)$ defines the *second fundamental form h* of $M$.

**Theorem 5.1.** $\nabla$ *is the Levi-Civita connection with respect to the induced Riemannian metric $g$.*

*Proof.* Since $\overline{\nabla}$ is the Levi-Civita connection with respect to $\bar{g}$, the torsion tensor $\overline{T}$ vanishes identically and therefore

$$\overline{T}(\imath X, \imath Y) = \overline{\nabla}_X \imath Y - \overline{\nabla}_Y \imath X - [\imath X, \imath Y] = 0.$$

Using relation (5.1), we compute

$$\imath \nabla_X Y + h(X,Y) - \imath \nabla_Y X - h(Y,X) - \imath[X,Y] = 0. \tag{5.2}$$

Considering the tangential part of relation (5.2), we conclude

$$\nabla_X Y - \nabla_Y X - [X,Y] = T(X,Y) = 0,$$

which implies that $\nabla$ is torsion-free. From the normal part of relation (5.2), we deduce

$$h(X,Y) = h(Y,X). \tag{5.3}$$

We now prove that $\nabla$ is a metric connection. Since $\overline{\nabla}$ is a metric connection of $\overline{M}$, we have

$$X(g(Y,Z)) = X(\bar{g}(\imath Y, \imath Z)) = \bar{g}(\overline{\nabla}_X \imath Y, \imath Z) + \bar{g}(\imath Y, \overline{\nabla}_X \imath Z). \tag{5.4}$$

On the other hand,

$$X(g(Y,Z)) = (\nabla_X g)(Y,Z) + g(\nabla_X Y, Z) + g(Y, \nabla_X Z)$$
$$= (\nabla_X g)(Y,Z) + \overline{g}(\imath \nabla_X Y, \imath Z) + \overline{g}(\imath Y, \imath \nabla_X Z)$$
$$= (\nabla_X g)(Y,Z) + \overline{g}(\overline{\nabla}_X \imath Y, \imath Z) + \overline{g}(\imath Y, \overline{\nabla}_X \imath Z). \qquad (5.5)$$

Comparing relations (5.4) and (5.5), we conclude $(\nabla_X g)(Y,Z) = 0$ and consequently $\nabla_X g = 0$. Thus, $\nabla$ is a metric connection, which completes the proof. $\qquad\square$

Let $\xi$ be a normal vector field on $M$. Then $\overline{\nabla}_X \xi$ splits into the tangential part and the normal part, that is, the following *Weingarten formula* holds:

$$\overline{\nabla}_X \xi = -\imath A_\xi X + D_X \xi. \qquad (5.6)$$

$A_\xi$ is called the *shape operator* with respect to the normal vector field $\xi$. It is easy to check that $A_\xi$ is a linear mapping from the tangent bundle $T(M)$ into itself and that $D$ defines a connection on the normal bundle $T^\perp(M)$. We call $D$ the *normal connection* of $M$ in $\overline{M}$.

Differentiating covariantly $\overline{g}(\imath Y, \xi) = 0$, we obtain

$$\overline{g}(\overline{\nabla}_X \imath Y, \xi) + \overline{g}(\imath Y, \overline{\nabla}_X \xi) = 0,$$

from which it follows

$$\overline{g}(\imath \nabla_X Y + h(X,Y), \xi) + \overline{g}(\imath Y, -\imath A_\xi X + D_X \xi) = 0.$$

Thus, we have

$$g(A_\xi X, Y) = \overline{g}(h(X,Y), \xi). \qquad (5.7)$$

Using (5.3) and (5.7), we conclude that the shape operator is symmetric.

Let $\xi_1, \ldots, \xi_p$ be an orthonormal frame of $T^\perp(M)$ and denote $A_{\xi_a}$ by $A_a$. Then the Weingarten formula (5.6) can be written as

$$\overline{\nabla}_X \xi_a = -\imath A_a X + D_X \xi_a, \quad D_X \xi_a = \sum_{b=1}^{p} s_{ab}(X)\xi_b, \qquad (5.8)$$

where the $s_{ab}$ are called the coefficients of the *third fundamental form* of $M$ in $\overline{M}$. For simplicity, we sometimes suppress the explicit dependence on $X$ in the notation. The coefficients of the third fundamental form satisfy

$$s_{ab} + s_{ba} = 0. \qquad (5.9)$$

Namely, using (5.8), we compute

$$X\overline{g}(\xi_a, \xi_b) = \overline{g}(\overline{\nabla}_X \xi_a, \xi_b) + \overline{g}(\xi_a, \overline{\nabla}_X \xi_b)$$
$$= \overline{g}(\sum_{c=1}^{p} s_{ac}(X)\xi_c, \xi_b) + \overline{g}(\xi_a, \sum_{c=1}^{p} s_{bc}(X)\xi_c)$$
$$= s_{ab}(X) + s_{ba}(X).$$

On the other hand, $X\overline{g}(\xi_a, \xi_b) = X\delta_{ab} = 0$. Comparing the last two equations, we obtain (5.9).

Particularly, in the case when the difference of the dimension of $\overline{M}$ and $M$ is two, we use the notation

$$s_{12} = -s_{21} = s. \tag{5.10}$$

**Definition 5.3.** A submanifold $M$ of $\overline{M}$ is called a *totally geodesic submanifold* of $M$ if for every geodesic $\gamma(s)$ of $M$, curve $\imath\gamma(s)$ is a geodesic of $\overline{M}$.

**Theorem 5.2.** *A submanifold $M$ is totally geodesic if and only if the second fundamental form $h$ vanishes identically.*

*Proof.* Let $\gamma(s)$ be a geodesic of $M$, i.e., let $\nabla_{\dot\gamma}\dot\gamma = 0$. Therefore, using relation (5.1), it follows

$$\overline{\nabla}_{\dot\gamma}\imath\dot\gamma = \imath\nabla_{\dot\gamma}\dot\gamma + h(\dot\gamma, \dot\gamma) = h(\dot\gamma, \dot\gamma).$$

Consequently, if the second fundamental form $h$ vanishes identically, $M$ is totally geodesic. Conversely, if $M$ is totally geodesic, it follows $h(\dot\gamma, \dot\gamma) = 0$. Since through any point $x \in M$, for any $X \in T_x(M)$, there exists a geodesic $\gamma(s)$ whose tangent vector at $x$ is $X_x$, we deduce $h(X, X) = 0$, for any $X$. Therefore, $h(X + Y, X + Y) = 0$ and using relation (5.3), we conclude $h(X, Y) = 0$, which completes the proof. $\square$

Theorem 5.2 and relation (5.7) imply

**Corollary 5.1.** *$M$ is a totally geodesic submanifold if and only if relation $A_\xi = 0$ holds for any normal vector field $\xi$ of $M$. Particularly, $M$ is totally geodesic if and only if $A_1 = \cdots = A_p = 0$ for an orthonormal frame field $\xi_1, \ldots, \xi_p$ of $T^\perp(M)$.*

We now consider how the shape operators $A_a$ and the third fundamental form $s_{ab}$ change when we choose another orthonormal frame field $T^\perp(M)$. Let

$$\xi'_a = \sum_{b=1}^{p} T_a^b \xi_b, \quad a = 1, \ldots, p. \tag{5.11}$$

Since $\xi'_1, \ldots, \xi'_p$ are orthonormal, we conclude $T_a^b(x) \in SO(p)$, at any point $x \in M$.

Now, let us compute $\overline{\nabla}_X \xi'_a$. First, using relations (5.8) and (5.11), we obtain

$$\overline{\nabla}_X \xi'_a = -\imath A'_a X + \sum_{c=1}^{p} s'_{ac}(X)\xi'_c$$

$$= -\imath A'_a X + \sum_{b,c=1}^{p} s'_{ac}(X) T_c^b \xi_b. \tag{5.12}$$

On the other hand, we have

$$
\overline{\nabla}_X \xi_a' = \sum_{b=1}^{p} \left\{ (XT_a^b)\xi_b + T_a^b \overline{\nabla}_X \xi_b \right\}
$$

$$
= \sum_{b=1}^{p} \left\{ (XT_a^b)\xi_b + T_a^b \left( -\imath A_b X + \sum_{c=1}^{p} s_{bc}(X)\xi_c \right) \right\}
$$

$$
= -\imath \sum_{b=1}^{p} T_a^b A_b X + \sum_{b=1}^{p} \left\{ XT_a^b + \sum_{c=1}^{p} T_a^c s_{cb}(X) \right\} \xi_b. \qquad (5.13)
$$

Comparing relations (5.13) and (5.12), we conclude

$$
A_a' = \sum_{b=1}^{p} T_a^b A_b, \qquad (5.14)
$$

$$
\sum_{c=1}^{p} s_{ac}'(X)T_c^b = XT_a^b + \sum_{c=1}^{p} T_a^c s_{cb}(X). \qquad (5.15)
$$

**Definition 5.4.** The vector field defined by

$$
\mu = \frac{1}{n} \sum_{a=1}^{p} (\text{trace } A_a)\xi_a
$$

is called the *mean curvature vector field* of $M$.

**Theorem 5.3.** *The mean curvature vector field $\mu$ is independent of the choice of orthonormal basis $\xi_1, \ldots, \xi_p$.*

*Proof.* Let $\xi_a'$, $a = 1, \ldots, p$ be another orthonormal frame field of $T^\perp(M)$. Then $\xi_a' = \sum_{b=1}^{p} T_a^b \xi_b$, for $T_a^b \in SO(p)$. Therefore, making use of (5.14), we obtain

$$
n\mu' = \sum_{a=1}^{p} (\text{trace } A_a')\xi_a' = \sum_{a,b,c=1}^{p} (T_a^b \, \text{trace } A_b)T_a^c \xi_c
$$

$$
= \sum_{a=1}^{p} (\text{trace } A_a)\xi_a = n\mu,
$$

which completes the proof. $\qquad \square$

**Definition 5.5.** The length $|\mu|$ of the mean curvature vector field $\mu$ is called the *mean curvature of the submanifold $M$*.

**Proposition 5.1.** *Let $x \in M$ be such a point that $\mu(x) \neq 0$. At $x$ we choose an orthonormal basis $\xi_1, \ldots, \xi_p$ of $T_x^\perp(M)$ in such a way that $\xi_1$ is in the direction of the mean curvature vector field $\mu$. Then*

$$
n|\mu| = \text{trace } A_1, \qquad \text{trace } A_a = 0 \quad \text{for } a = 2, \ldots, p.
$$

*Proof.* Definition 5.4 and Theorem 5.3 imply

$$n\,\mu = n\,|\mu|\,\xi_1 = \sum_{a=1}^{p}(\text{trace } A_a)\xi_a,$$

from which it follows

$$(\text{trace } A_1 - n|\mu|)\xi_1 + \sum_{a=2}^{p}(\text{trace } A_a)\xi_a = 0.$$

Since $\xi_1, \ldots, \xi_p$ are orthonormal, we conclude that $n\,|\mu| = \text{trace } A_1$ and $\text{trace } A_a = 0$ for $a = 2, \ldots, p$. $\qquad\square$

**Proposition 5.2.** *The mean curvature vector field $\mu$ is parallel with respect to the normal connection if and only if*

$$X\,(\text{trace} A_a) + \sum_{b=1}^{p}(\text{trace} A_b)\,s_{ba}(X) = 0, \quad a = 1, \ldots, p. \qquad (5.16)$$

*Proof.* Definition 5.4 and relation (5.8) imply

$$n\overline{\nabla}_X\mu = \sum_{a=1}^{p}\overline{\nabla}_X\left((\text{trace} A_a)\xi_a\right)$$

$$= \sum_{a=1}^{p}\left\{(X(\text{trace} A_a))\xi_a + (\text{trace} A_a)\left(-\imath A_a X + \sum_{b=1}^{p}s_{ab}(X)\xi_b\right)\right\}$$

$$= -\imath\sum_{a=1}^{p}(\text{trace} A_a)\,A_a X + \sum_{a=1}^{p}\left\{X(\text{trace} A_a) + \sum_{b=1}^{p}(\text{trace} A_b)s_{ba}(X)\right\}\xi_a.$$

Thus we have

$$n\,D_X\mu = \sum_{a=1}^{p}\left\{X\,(\text{trace } A_a) + \sum_{b=1}^{p}(\text{trace } A_b)\,s_{ba}(X)\right\}\xi_a,$$

which establishes the formula (5.16). $\qquad\square$

**Proposition 5.3.** *If $\mu$ is parallel with respect to the normal connection, then the mean curvature is constant.*

*Proof.* From Proposition 5.2, if $\mu$ is parallel with respect to the normal connection, it follows

$$X|\mu|^2 = \frac{2}{n^2}\sum_{a=1}^{p}(X(\text{trace } A_a))\,(\text{trace } A_a)$$

$$= -\frac{2}{n^2}\sum_{a,b=1}^{p}(\text{trace } A_a)(\text{trace } A_b)s_{ba}(X) = 0,$$

since relation (5.9) states that the coefficients of the third fundamental form are skew-symmetric. $\qquad\square$

**Definition 5.6.** If the mean curvature vector field $\mu$ vanishes at a point $x \in M$, the point $x$ is called a minimal point and if $\mu$ vanishes identically on $M$, submanifold $M$ is called a *minimal submanifold* of $\overline{M}$.

From Definition 5.6 we easily deduce

**Proposition 5.4.** *The submanifold $M$ is minimal if and only if* trace $A_a = 0$, $a = 1, \ldots, p$.

Let $\xi_1, \ldots, \xi_p$ be mutually orthonormal normal vector fields of $M$ and let $A_a X = \rho_a X$, $a = 1, \ldots, p$ for any $X \in T(M)$, for the shape operators $A_a$. From (5.14) it may be concluded that, if we take another orthonormal basis $\xi_1', \ldots, \xi_p'$, the corresponding shape operators $A_a'$ satisfy $A_a' X = \rho_a' X$. Therefore, we can give the following

**Definition 5.7.** If the shape operator $A_a$ satisfies $A_a X = \rho_a X$ for $a = 1, \ldots, p$, the submanifold $M$ is called a *totally umbilical submanifold*.

Now we study the case when a local orthonormal normal frame field $\xi_1', \ldots, \xi_p'$ can be chosen in such a way that the third fundamental form vanishes, namely, $s_{ab}' = 0$ for $a, b = 1, \ldots, p$. Using relation (5.15), this is equivalent to finding $T_a^b \in SO(p)$ which satisfy

$$dT_a^b + \sum_{c=1}^{p} T_a^c s_{cb} = 0. \tag{5.17}$$

By the well-known Poincaré Lemma, it follows that the existence of a solution of the equation

$$d \sum_{c=1}^{p} T_a^c s_{cb} = 0 \tag{5.18}$$

guarantees the existence of such $T_a^b$. Since equation (5.18) is equivalent to

$$\sum_{c=1}^{p} dT_a^c \wedge s_{cb} + \sum_{c=1}^{p} T_a^c \, ds_{cb} = 0, \tag{5.19}$$

using equation (5.17), we compute

$$\sum_{d=1}^{p} T_a^d \left( ds_{db} - \sum_{c=1}^{p} s_{dc} \wedge s_{cb} \right) = 0. \tag{5.20}$$

As $T_a^d \in SO(p)$, using (5.20), we conclude that we can choose a local orthonormal frame field of $T^{\perp}(M)$ in such a way that the third fundamental form vanishes, if the following relation holds:

$$ds_{db} - \sum_{c=1}^{p} s_{dc} \wedge s_{cb} = 0, \quad \text{for} \quad d, b = 1, \ldots, p. \tag{5.21}$$

Now, let $\overline{R}$ denote the curvature tensor of $\overline{M}$. Then, using the Gauss formula (5.1) and the Weingarten formula (5.8), for $X, Y \in T(M)$, we obtain

$$\overline{R}(\imath X, \imath Y)\imath Z = \overline{\nabla}_X \overline{\nabla}_Y \imath Z - \overline{\nabla}_Y \overline{\nabla}_X \imath Z - \overline{\nabla}_{[X,Y]} \imath Z$$

$$= \overline{\nabla}_X \left( \imath \nabla_Y Z + \sum_{a=1}^{p} g(A_a Y, Z)\xi_a \right) - \overline{\nabla}_Y \Bigg( \imath \nabla_X Z$$

$$+ \sum_{a=1}^{p} g(A_a X, Z)\xi_a \Bigg) - \imath \nabla_{[X,Y]} Z - \sum_{a=1}^{p} g(A_a[X,Y], Z)$$

$$= \imath \nabla_X \nabla_Y Z + \sum_b g(A_b X, \nabla_Y Z)\xi_b$$

$$+ \sum_a \{ g((\nabla_X A_a)Y, Z) + g(A_a \nabla_X Y, Z) + g(A_a Y, \nabla_X Z) \} \xi_a$$

$$+ \sum_a g(A_a Y, Z)(-\imath A_a X + \sum_b s_{ab}(X)\xi_b)$$

$$- \imath \nabla_Y \nabla_X Z - \sum_b g(A_a Y, \nabla_X Z)\xi_b$$

$$- \sum_a \{ (g(\nabla_Y A_a)X, Z) + g(A_a \nabla_Y X, Z) + g(A_a X, \nabla_Y Z) \} \xi_a$$

$$- \sum_a g(A_a X, Z)(-\imath A_a Y + \sum_b s_{ab}(Y)\xi_b)$$

$$- \imath \nabla_{[X,Y]} Z - \sum_a g(A_a[X,Y], Z)\xi_a$$

$$= \imath \Bigg\{ \nabla_X \nabla_Y Z - \nabla_Y \nabla_X Z - \nabla_{[X,Y]} Z$$

$$- \sum_a (g(A_a Y, Z)A_a X - g(A_a X, Z)A_a Y) \Bigg\}$$

$$+ \sum_a \Bigg\{ g((\nabla_X A_a)Y, Z) + \sum_b s_{ba}(X)g(A_b Y, Z)$$

$$- g((\nabla_Y A_a)X, Z) - \sum_b s_{ba}(Y)g(A_b X, Z) \Bigg\} \xi_a$$

$$= \imath \Bigg\{ R(X, Y)Z - \sum_a (g(A_a Y, Z)A_a X - g(A_a X, Z)A_a Y) \Bigg\}$$

$$+ \sum_a \Bigg\{ g((\nabla_X A_a)Y - (\nabla_Y A_a)X, Z)$$

$$+ \sum_b s_{ba}(X)g(A_b Y, Z) - s_{ba}(Y)g(A_b X, Z) \Bigg\} \xi_a.$$

Thus we have the following *Gauss equation:*

$$\bar{g}(\bar{R}(\imath X, \imath Y)\imath Z, \imath W) = g(R(X,Y)Z,W) - \sum_{a=1}^{p} \Big\{ g(A_a Y, Z)g(A_a X, W)$$
$$- g(A_a X, Z)g(A_a Y, W) \Big\}, \tag{5.22}$$

and *Codazzi equation*

$$\bar{g}(\bar{R}(\imath X, \imath Y)\imath Z, \xi_a) = g\Big( (\nabla_X A_a)Y - (\nabla_Y A_a)X, Z \Big)$$
$$+ \sum_{b=1}^{p} \Big\{ s_{ba}(X)g(A_b Y, Z) - s_{ba}(Y)g(A_b X, Z) \Big\}. \tag{5.23}$$

After computing $\bar{R}(\imath X, \imath Y)\xi_a$ in the same way, we obtain

$$\bar{R}(\imath X, \imath Y)\xi_a = \bar{\nabla}_X \bar{\nabla}_Y \xi_a - \bar{\nabla}_Y \bar{\nabla}_X \xi_a - \bar{\nabla}_{[X,Y]}\xi_a$$
$$= \imath \Big\{ -(\nabla_X A_a)Y + (\nabla_Y A_a)X - \sum_b (s_{ab}(Y)A_b X - s_{ab}(X)A_b Y) \Big\}$$
$$+ \sum_b \Big\{ (\nabla_X s_{ab})(Y) - (\nabla_Y s_{ab})(X) - g((A_a A_b - A_b A_a)X, Y)$$
$$+ \sum_c [s_{ac}(Y)s_{cb}(X) - s_{ac}(X)s_{cb}(Y)] \Big\} \xi_b,$$

and we deduce the following *Ricci-Kühne equation:*

$$\bar{g}(\bar{R}(\imath X, \imath Y)\xi_a, \xi_b) = g\Big( (A_b A_a - A_a A_b)X, Y \Big) + (\nabla_X s_{ab})(Y) - (\nabla_Y s_{ab})(X)$$
$$+ \sum_c \Big\{ s_{ac}(Y)s_{cb}(X) - s_{ac}(X)s_{cb}(Y) \Big\}. \tag{5.24}$$

We define the *normal curvature* $R^{\perp}$ of $M$ in $\bar{M}$ in the following way:

$$R^{\perp}(X,Y)\xi_a = D_X D_Y \xi_a - D_Y D_X \xi_a - D_{[X,Y]}\xi_a.$$

Then, we compute

$$R^{\perp}(X,Y)\xi_a = \left(\overline{\nabla}_X\left(\sum_b s_{ab}(Y)\xi_b\right)\right)^{\perp} - \left(\overline{\nabla}_Y\left(\sum_b s_{ab}(X)\xi_b\right)\right)^{\perp}$$
$$- \sum_b s_{ab}\left([X,Y]\right)\xi_b$$
$$= \sum_b \left\{X(s_{ab}(Y))\xi_b + \sum_c s_{ab}(Y)s_{bc}(X)\xi_c - Y\left(s_{ab}(X)\right)\xi_b\right.$$
$$\left. - \sum_c s_{ab}(X)s_{bc}(Y)\xi_c\right\} - \sum_b s_{ab}\left([X,Y]\right)\xi_b$$
$$= \sum_{b=1}^{p}\left\{X\left(s_{ab}(Y)\right) - Y\left(s_{ab}(X)\right) - s_{ab}\left([X,Y]\right)\right.$$
$$\left. + \sum_{c=1}^{p}\left(s_{ac}(Y)s_{cb}(X) - s_{ac}(X)s_{cb}(Y)\right)\right\}\xi_b$$
$$= \sum_{b=1}^{p}\left\{ds_{ab}(X,Y) - \sum_{c=1}^{p} s_{ac}\wedge s_{cb}(X,Y)\right\}\xi_b. \tag{5.25}$$

Hence, using (5.25), we conclude

$$\overline{g}(R^{\perp}(X,Y)\xi_a,\xi_b) = ds_{ab}(X,Y) - \sum_{c=1}^{p} s_{ac}\wedge s_{cb}(X,Y). \tag{5.26}$$

Thus, relations (5.24), (4.4) and (5.26) imply *Ricci equation*

$$\overline{g}(\overline{R}(\imath X,\imath Y)\xi_a,\xi_b) = g\left([A_b,A_a]X,Y\right) + \overline{g}(R^{\perp}(X,Y)\xi_a,\xi_b). \tag{5.27}$$

If the normal curvature $R^{\perp}$ of $M$ in $\overline{M}$ vanishes identically, we say that the normal connection of $M$ is flat. Using relations (5.21) and (5.26), we prove the following meaning of the normal connection:

**Proposition 5.5.** *If the normal curvature $R^{\perp}$ vanishes identically, we can choose an orthonormal frame field $\xi_a$, $a = 1,\dots,p$ of $T^{\perp}(M)$ in such a way that the third fundamental form vanishes.*

# 6

## Submanifolds of a Euclidean space

In this section, we give characterizations of typical submanifolds of Euclidean space. First of all, we prove the following.

**Theorem 6.1.** *An $n$-dimensional totally geodesic submanifold $M$ of $(n+p)$-dimensional Euclidean space $\mathbf{E}^{n+p}$ is an open submanifold of $n$-dimensional Euclidean space. If $M$ is complete, then $M$ is an $n$-dimensional Euclidean space.*

*Proof.* Since the ambient manifold is a Euclidean space, the Ricci-Kühne equation (5.27) implies $R^\perp(X,Y)\xi_a = 0$, that is, the normal curvature vanishes identically. According to Proposition 5.5, we can choose orthonormal normal vector fields $\xi_1,\ldots,\xi_p$ to $M$ in such a way that the corresponding third fundamental form will vanish. Therefore, $\overline{\nabla}_X\xi_a = 0$, since $M$ is a totally geodesic submanifold.

Now, let us define in a neighborhood $U(x)$ of $x \in M$, $p$ functions $f_a$, $a = 1,\ldots,p$ by $f_a = \langle \mathbf{x}, \xi_a \rangle$, where $\mathbf{x}$ denotes the position vector field of $x \in M$ and $\langle,\rangle$ the Euclidean metric of the ambient manifold $\mathbf{E}^{n+p}$. Then

$$X f_a = X\langle \mathbf{x}, \xi_a \rangle = \langle \imath X, \xi_a \rangle + \langle \mathbf{x}, \overline{\nabla}_X \xi_a \rangle = 0,$$

holds for any $X \in T(M)$, which means that $f_a = constant$ for $a = 1,\ldots,p$. Thus, for $\mathbf{x} = \sum_{i=1}^{n+p} x^i e_i$, $\xi_a = \sum_{i=1}^{n+p} \xi_a^i e_i$, we compute

$$\langle \mathbf{x}, \xi_a \rangle = x^1 \xi_a^1 + \cdots + x^{n+p} \xi_a^{n+p} = c_a, \quad a = 1,\ldots,p.$$

This shows that $U(x)$ lies in the intersection of $p$ hyperplanes whose normal vectors are linearly independent, that is, there exists an $n$-dimensional Euclidean space $\mathbf{E}^n$ such that $U(x) \subset \mathbf{E}^n$. However, since $U(x)$ is an open subset of $\mathbf{E}^n$, $U(x)$ is $n$-dimensional. Therefore, $M$ is an open subset of $\mathbf{E}^n$. Particularly, if $M$ is complete, $M$ is an $n$-dimensional Euclidean space. This completes the proof. $\square$

M. Djorić, M. Okumura, *CR Submanifolds of Complex Projective Space*,
Developments in Mathematics 19, DOI 10.1007/978-1-4419-0434-8_6,
© Springer Science+Business Media, LLC 2010

Next we consider a totally umbilical submanifold of Euclidean space. In the following, we assume that a totally umbilical submanifold means it is not a totally geodesic submanifold.

**Theorem 6.2.** *An $n$-dimensional totally umbilical submanifold $M$ of $(n+p)$-dimensional Euclidean space $\mathbf{E}^{n+p}$ is an open submanifold of $n$-dimensional sphere $\mathbf{S}^n$ and if $M$ is complete, $M$ is an $n$-dimensional sphere.*

*Proof.* Since $M$ is not totally geodesic, we can choose orthonormal normal fields $\xi_1, \xi'_2, \ldots, \xi'_p$ in such a way that $\xi_1$ is a unit vector field in the direction of the mean curvature vector field $\mu$. Proposition 5.1 now implies trace $A'_a = 0$ for $a = 2, \ldots, p$, where $A'_a$ denotes the shape operator with respect to the normal $\xi'_a$. Analysis similar to that in the proof of Theorem 6.1, using the Ricci-Kühne equation (5.27), shows that we can choose orthonormal fields $\xi_2, \ldots, \xi_p$ normal to $M$, such that the corresponding third fundamental form will vanish. Hence we have

$$\overline{\nabla}_X \imath Y = \imath \nabla_X Y + g(A_1 X, Y)\xi_1. \tag{6.1}$$

Moreover, the Codazzi equation (5.23) reduces to $(X\rho_1)Y = (Y\rho_1)X$. Since $X$ and $Y$ are linearly independent, we conclude that $\rho_1$ is constant.

Now we define $p - 1$ functions $f_a$ by $f_a = \langle \mathbf{x}, \xi_a \rangle$, $a = 2, \ldots, p$. Then we obtain that the $f_a$ are constant and in entirely the same manner as in the proof of Theorem 6.1, we deduce that $M$ lies in the intersection of $p - 1$ hyperplanes whose normal vectors are linearly independent, that is, there exists an $(n + 1)$-dimensional Euclidean space $\mathbf{E}^{n+1}$ such that $M \subset \mathbf{E}^{n+1}$.

Further, we denote by $j$ and $j'$ the immersion $M \to \mathbf{E}^{n+1}$ and the totally geodesic immersion $\mathbf{E}^{n+1} \to \mathbf{E}^{n+p}$, respectively. Then $\imath = j' \circ j$ and

$$\begin{aligned}
\overline{\nabla}_X \imath Y &= \overline{\nabla}_X j' \circ jY = j' \nabla'_X jY \\
&= j'(j\nabla_X Y + g(A'X, Y)\xi') = \imath \nabla_X Y + g(A'X, Y)j'\xi'. \tag{6.2}
\end{aligned}$$

Comparing relations (6.1) and (6.2), we obtain $A'X = A_1 X = \rho_1 X$.

As $M$ is a hypersurface of $\mathbf{E}^{n+1}$, the position vector field $\mathbf{x}$ of $M$ satisfies

$$X\left(\mathbf{x} + \frac{1}{\rho_1}\xi'\right) = 0, \qquad |\mathbf{x} - P| = \frac{1}{|\rho_1|},$$

for any $X \in T(M)$, where $P = \mathbf{x} + \frac{1}{\rho_1}\xi'$. This shows that $P$ is a fixed point, in $\mathbf{E}^{n+1}$, for $\mathbf{x} \in M$ and that any point of $M$ has constant distance $\frac{1}{|\rho_1|}$ from the fixed point $P$. Hence $M$ lies on a sphere of radius $\frac{1}{|\rho_1|}$ in $\mathbf{E}^{n+1}$ and the theorem follows.   $\square$

# 7

# Submanifolds of a complex manifold

Let $\overline{M}$ be a complex manifold with the natural almost complex structure $J$ and let $M$ be a real submanifold of $\overline{M}$.

**Proposition 7.1.** *For any point $x \in M$, let $S_x(M)$ be a subspace of the tangent space $T_x(M)$. Then $S_x(M) \cap JS_x(M)$ is a $J$-invariant subspace of $T_x(M)$.*

*Proof.* Let us suppose $X \in S_x(M) \cap JS_x(M)$. Then $X \in S_x(M)$, implies $JX \in JS_x(M)$. On the other hand, $X \in JS_x(M)$ implies the existence of $Y \in S_x(M)$ such that $JY = X$. Hence, $JX = J^2Y = -Y \in S_x(M)$. Thus, $JX \in S_x(M) \cap JS_x(M)$, which shows that $S_x(M) \cap JS_x(M)$ is $J$-invariant subspace of $T_x(M)$. $\qquad\square$

**Definition 7.1.** We call $H_x(M) = JT_x(M) \cap T_x(M)$ the *holomorphic tangent space* of $M$.

**Proposition 7.2.** *$H_x(M)$ is the maximal $J$-invariant subspace of $T_x(M)$.*

*Proof.* According to Proposition 7.1, $H_x(M)$ is a $J$-invariant subspace of $T_x(M)$. To prove that $H_x(M)$ is a maximal $J$-invariant subspace, let $T'_x(M)$ be a $J$-invariant subspace of $T_x(M)$. Then we have $JT'_x(M) \subset T'_x(M)$. For any $X \in T'_x(M) \subset T_x(M)$, it follows $JX \in JT'_x(M) \subset T'_x(M)$. We denote $JX = Y$, then $-X = J^2X = JY \in JT'_x(M) \subset JT_x(M)$. Hence $X \in JT_x(M)$ and consequently $X \in H_x(M)$. This shows that $T'_x(M) \subset H_x(M)$, which completes the proof. $\qquad\square$

The totally real part of $T_x(M)$ is $R_x(M) = T_x(M)/H_x(M)$.

**Proposition 7.3.** *$JR_x(M) \cap R_x(M) = \{0\}$.*

*Proof.* Since $H_x(M)$ is the maximal $J$-invariant subspace of $T_x(M)$, $JR_x(M) \cap R_x(M) \subset H_x(M)$. Hence if $X \in JR_x(M) \cap R_x(M)$, then $X \in R_x(M) \cap H_x(M)$ which implies that $X = 0$. Thus, $JR_x(M) \cap R_x(M) = \{0\}$. $\qquad\square$

From the above it also follows

M. Djorić, M. Okumura, *CR Submanifolds of Complex Projective Space*, Developments in Mathematics 19, DOI 10.1007/978-1-4419-0434-8_7, © Springer Science+Business Media, LLC 2010

**Proposition 7.4.** $T_x(M) = H_x(M) \oplus R_x(M)$.

**Proposition 7.5.** *Let $M$ be an $n$-dimensional submanifold of real $(n + p)$-dimensional complex manifold $(\overline{M}, J)$. Then we have*

$$n - p \leq \dim_{\mathbf{R}} H_x(M) \leq n. \tag{7.1}$$

*Proof.* $H_x(M) \subset T_x(M)$ implies that $\dim_{\mathbf{R}} H_x(M) \leq \dim T_x(M) = n$. On the other hand, $T_x(M) + JT_x(M) \subset T_{i(x)}(\overline{M})$ implies that

$$\dim_{\mathbf{R}} T_{i(x)}(\overline{M}) \geq \dim T_x(M) + \dim JT_x(M) - \dim_{\mathbf{R}} H_x(M),$$

from which

$$n + p \geq 2n - \dim_{\mathbf{R}} H_x(M).$$

Hence we have $\dim_{\mathbf{R}} H_x(M) \geq n - p$, which completes the proof.    □

From Proposition 7.5, it may be concluded that $\dim_{\mathbf{R}} H_x(M)$ is an even number between $n - p$ and $n$. Therefore, under the above assumptions, we have

**Corollary 7.1.** $0 \leq \dim_{\mathbf{R}} R_x(M) \leq p$.

*Proof.* Since $\dim_{\mathbf{R}} R_x(M) = \dim_{\mathbf{R}} T_x(M) - \dim_{\mathbf{R}} H_x(M)$, using Proposition 7.5, we establish the formula.    □

*Example 7.1.* Let

$$M = \{z \in \mathbf{C}^n \mid |z| = 1, Im z^n = 0\}$$

$$= \{(x^1, y^1, \ldots, x^n, y^n) \in \mathbf{R}^{2n} \mid \sum_{i=1}^{n}((x^i)^2 + (y^i)^2) = 1, y^n = 0\}.$$

Then $\dim M = 2n - 2$ and $p = 2$ and $\frac{\partial}{\partial y^n}$ is normal to $M$. From Proposition 7.5, it follows

$$2n - 4 \leq \dim_{\mathbf{R}} H_x(M) \leq 2n - 2.$$

Let $p_1$ be the point of $M$, represented by

$$z^1 = z^2 = \cdots = z^{n-2} = 0, \ z^{n-1} = 1, \ z^n = 0.$$

As a point of $\mathbf{R}^{2n}$, $p_1$ is represented by

$$x^1 = y^1 = \cdots = x^{n-2} = y^{n-2} = 0, \ x^{n-1} = 1, \ y^{n-1} = x^n = y^n = 0.$$

Therefore, $\frac{\partial}{\partial x^{n-1}}$ is a normal vector to $M$ at $p_1$. Hence

$$T_{p_1}(M) = \text{span} \left\{ \frac{\partial}{\partial x^1}, \frac{\partial}{\partial y^1}, \ldots, \frac{\partial}{\partial x^{n-2}}, \frac{\partial}{\partial y^{n-2}}, \frac{\partial}{\partial y^{n-1}}, \frac{\partial}{\partial x^n} \right\}$$

and $J(\frac{\partial}{\partial x^n}) = \frac{\partial}{\partial y^n}$, $J(\frac{\partial}{\partial y^{n-1}}) = -\frac{\partial}{\partial x^{n-1}}$ are orthogonal to $T_{p_1}(M)$. Therefore

$$R_{p_1}(M) = \text{span} \left\{ \frac{\partial}{\partial y^{n-1}}, \frac{\partial}{\partial x^n} \right\},$$

$$H_{p_1}(M) = \text{span} \left\{ \frac{\partial}{\partial x^1}, \frac{\partial}{\partial y^1}, \ldots, \frac{\partial}{\partial x^{n-2}}, \frac{\partial}{\partial y^{n-2}} \right\}.$$

This shows that $\dim_{\mathbf{R}} H_{p_1}(M) = 2n - 4$.

Next, we take the point $p_2 \in M$ represented by

$$z^1 = 0, \ldots, z^{n-1} = 0, z^n = 1.$$

As a point of $R^{2n}$, $p_2$ is represented by

$$x^1 = y^1 = \cdots = x^{n-1} = y^{n-1} = 0, x^n = 1, y^n = 0.$$

Then $\frac{\partial}{\partial x^n}, \frac{\partial}{\partial y^n}$ are normal vectors to $M$, at $p_2$ and $JT_{p_2}(M) = T_{p_2}(M)$, since $J(\frac{\partial}{\partial x^i}) = \frac{\partial}{\partial y^i}, J(\frac{\partial}{\partial y^i}) = -\frac{\partial}{\partial x^i}$. Hence $H_{p_2}(M) = T_{p_2}(M)$ and $\dim_{\mathbf{R}} H_{p_2}(M) = 2n - 2$.

From this example we conclude that, in general, the dimension of $H_p(M)$ varies depending on the point $p \in M$.                                              ◇

Now we give the definition of a CR submanifold.

**Definition 7.2.** [41]   If $H_x(M)$ has constant dimension with respect to $x \in M$, the submanifold $M$ is called a *Cauchy-Riemann submanifold* or briefly CR submanifold and the constant complex dimension is called the *CR dimension* of $M$.

*Example 7.2.    J-invariant submanifolds.*

Let $(\overline{M}, J)$ be a complex manifold and $\imath : M \to \overline{M}$ be an embedding. If for any $x \in M$, the subspace $\imath T_x(M)$ is an invariant subspace of $T_{\imath(x)}(\overline{M})$ with respect to $J$, that is,

$$J\imath T_x(M) \subset \imath T_x(M),$$

the submanifold $M$ is called a *J-invariant submanifold*, or *invariant submanifold*, for short. Moreover, since $J$ is an isomorphism, we conclude

$$J\imath T_x(M) = \imath T_x(M).$$

Consequently,

$$H_x(M) = T_x(M), \quad \dim_{\mathbf{R}} H_x(M) = n$$

and $M$ is a CR submanifold.

**Theorem 7.1.** *An invariant submanifold $M$ of a complex manifold $(\overline{M}, J)$ is a complex submanifold.*

*Proof.* Since $J\imath T_x(M) \subset \imath T(M)$, for any $X \in T_x(M)$, we may put

$$J\imath X = \imath J'X. \tag{7.2}$$

Then, $J' : T_x(M) \to T_x(M)$ is an isomorphism. Moreover,

$$-\imath X = J^2\imath X = J\imath J'X = \imath(J')^2 X$$

implies $(J')^2 X = -X$. Thus, $J'$ defines an almost complex structure on $M$.

Further, the Nijenhuis tensor $\overline{N}$ with respect to $J$ vanishes identically on $\overline{M}$, since $J$ is the almost complex structure induced from the complex structure of $\overline{M}$. Particularly for $X, Y \in T_x(M)$, it follows

$$
\begin{aligned}
\overline{N}(\imath X, \imath Y) &= J[\imath X, \imath Y] - [J\imath X, \imath Y] - [\imath X, J\imath Y] - J[J\imath X, J\imath Y] \\
&= J\imath[X,Y] - [\imath J'X, \imath Y] - [\imath X, \imath J'Y] - J[\imath J'X, \imath J'Y] \\
&= \imath\left(J'[X,Y] - [J'X,Y] - [X,J'Y] - J'[J'X,J'Y]\right) \\
&= \imath N(X,Y) = 0.
\end{aligned}
$$

This, together with relation (7.2) and Theorem 2.1, implies that $\imath$ is holomorphic and $(M, J')$ is a complex manifold, with the induced almost complex structure $J'$ from $J$ of $\overline{M}$. Consequently, $\dim_{\mathbf{R}} M$ is even.    □

*Example 7.3.    Real hypersurfaces.*

Since for a real hypersurface $M^n$ of a complex manifold from Proposition 7.5 it follows $\dim_{\mathbf{R}} H_x(M) = n - 1$, we conclude that $M$ is a CR submanifold.

*Example 7.4.    Totally real submanifolds.*

If $H_x(M) = \{0\}$ holds at every point $x \in M$, $M$ is called a *totally real submanifold*. We remark that for a totally real submanifold $M$ of $\overline{M}$, it follows $R_x(M) = T_x(M)$.

Using Proposition 7.5 we conclude that if $M$ is a totally real submanifold, we have

$$n - p \leq \dim_{\mathbf{R}} H_x(M) = 0.$$

Therefore $n \leq p$ and the following proposition follows.

**Proposition 7.6.** *The dimension of a totally real submanifold is less than or equal to the codimension of the submanifold in the ambient manifold.*

We now present one example of a totally real submanifold. Let

$$
\begin{aligned}
M &= \{x + \sqrt{-1}y \in \mathbf{C}^n \,|\, y = 0\} \\
&= \{(x^1, 0, x^2, 0, \ldots, x^n, 0) \,|\, x^i \in \mathbf{R}, \, i = 1, \ldots, n\}.
\end{aligned}
$$

Since in this case

$$T_x(M) = \text{span} \left\{ \frac{\partial}{\partial x^1}, \frac{\partial}{\partial x^2}, \cdots, \frac{\partial}{\partial x^n} \right\}$$

and $J(\frac{\partial}{\partial x^i}) = \frac{\partial}{\partial y^i}$, it follows $JT_x(M) \cap T_x(M) = \{0\}$, and therefore, $M$ is a totally real submanifold of $n$-dimensional complex space $\mathbf{C}^n$.    $\diamond$

*Example 7.5.* Let $(\overline{M}_1, J_1)$ and $(\overline{M}_2, J_2)$ be complex manifolds with almost complex structures $J_1$ and $J_2$. Then $\overline{M} = \overline{M}_1 \times \overline{M}_2$ is a complex manifold with respect to the almost complex structure $J = J_1 \oplus J_2$.

For a totally real submanifold $M_1$ of $\overline{M}_1$ and a complex submanifold $M_2$ of $\overline{M}_2$, the product manifold $M = M_1 \times M_2$ is a CR submanifold of $\overline{M}$ and $\dim(JT(M) \cap T(M)) = \dim(JT(M_2) \cap T(M_2)) = \dim M_2$.    $\diamond$

Now, let $M$ be a real hypersurface of a Kähler manifold. Then

$$\dim_{\mathbf{R}} R_x(M) = 1, \qquad JR_x(M) \perp R_x(M),$$

because of the skew symmetric property of $J$. On the other hand, for $X \in R_x(M)$, $Y \in H_x(M)$, we have $g(JX, Y) = -g(X, JY) = 0$ and hence $JR_x(M) \perp H_x(M)$. Consequently,

$$JR_x(M) \perp H_x(M) \oplus R_x(M) = T_x(M),$$

that is, $JR_x(M)$ is orthogonal to $T_x(M)$. However, for the higher codimension case $JR_x(M)$ is not always orthogonal to $T_x(M)$ and we provide a counterexample in the following.

*Example 7.6.* Let

$$M = \{ (z^1, z^2) \in \mathbf{C}^2 | \, Im\, z^1 = Re\, z^2, \, Im\, z^2 = 0 \},$$
$$= \{ (x^1, y^1, y^1, 0) \in \mathbf{R}^4 | \, x^1, y^1 \in \mathbf{R} \}$$

denote the submanifold of $\mathbf{C}^2$. Then, since

$$T_x(\mathbf{C}^2) = \text{span} \left\{ \frac{\partial}{\partial x^1}, \frac{\partial}{\partial y^1}, \frac{\partial}{\partial x^2}, \frac{\partial}{\partial y^2} \right\},$$

and since a point $x \in M$ can be described as the position vector, which can be expressed as a linear combination of basis vectors, it follows

$$x = x^1 \frac{\partial}{\partial x^1} + y^1 \frac{\partial}{\partial y^1} + y^1 \frac{\partial}{\partial x^2}.$$

Therefore,

$$T_x(M) = \text{span} \left\{ \frac{\partial}{\partial x^1}, \frac{\partial}{\partial y^1} + \frac{\partial}{\partial x^2} \right\}$$

and since

$$J\left(\frac{\partial}{\partial x^1}\right) = \frac{\partial}{\partial y^1},$$

$$J\left(\frac{\partial}{\partial y^1} + \frac{\partial}{\partial x^2}\right) = -\frac{\partial}{\partial x^1} + \frac{\partial}{\partial y^2},$$

it follows $H_x(M) = \{0\}$. However, $J\left(\frac{\partial}{\partial x^1}\right) = \frac{\partial}{\partial y^1}$ is not orthogonal to $T_x(M)$.

Let $(\overline{M}, J, \overline{g})$ be a Hermitian manifold. In [1] A. Bejancu gave another definition of CR submanifolds.

**Definition 7.3.** [1] $M$ is called a *CR submanifold* if there exists a pair of orthogonal complementary distributions $(\Delta, \Delta^\perp)$ of $T(M)$ such that

$$\text{for any} \quad x \in M, \quad J\Delta_x = \Delta_x \quad \text{and} \quad J\Delta_x^\perp \subset T_x(M)^\perp.$$

**Proposition 7.7.** *If $M$ is a CR submanifold in the sense of Definition 7.3, then $M$ is a CR submanifold in the sense of Definition 7.2.*

*Proof.* First we note that $\Delta_x \subset H_x(M)$, since from Proposition 7.2 we know that $H_x(M)$ is the maximal $J$-invariant subspace of $T_x(M)$. Further, if there exists $X \in H_x(M)$ such that $X \notin \Delta_x$, then

$$X = X_1 + X_2, \quad X_1 \in \Delta_x, \quad X_2 \in \Delta_x^\perp,$$

since $\Delta_x$ and $\Delta_x^\perp$ are mutually complement. Then it follows $JX = JX_1 + JX_2$ where $JX_2 \in T_x(M)^\perp$, contrary to $X \in H_x(M)$. Therefore, $\Delta_x = H_x(M)$. Finally, since $\Delta$ is a distribution, $dim\Delta_x$ is constant, which completes the proof. $\qquad\square$

If the CR dimension of $M^n$ is $\frac{n-1}{2}$, we call $M$ a *CR submanifold of maximal CR dimension*. In that case, let $e_1, \ldots, e_n$ be an orthonormal basis of $T_x(M)$ such that $e_1, \ldots, e_{n-1} \in H_x(M)$. Then $e_n \in T_x(M) \setminus H_x(M)$ and $\overline{g}(Je_n, \imath e_j) = -\overline{g}(\imath e_n, J\imath e_j) = 0$, $j = 1, \ldots, n-1$, since $J\imath e_j \in H_x(M)$. Therefore, $J\imath e_n \in T_x^\perp(M)$ and, using Proposition 7.7, we conclude

**Proposition 7.8.** *If $M$ is a CR submanifold of maximal CR dimension, then Definition 7.2 and Definition 7.3 are equivalent.*

On the other hand, as we show in the following, when the CR dimension of $M$ is less than $\frac{n-1}{2}$, the converse of Proposition 7.7 is false.

Let $M$ be a CR submanifold of CR dimension $\frac{n-2}{2}$. Choosing an orthonormal basis $e_1, e_2, \ldots, e_{n-2}, e_{n-1}, e_n$ of $T_x(M)$ in such a way that

$$e_1, e_2, \ldots, e_{n-2} \in JT_x(M) \cap T_x(M),$$

we can write

$$Je_i \in JT_x(M) \cap T_x(M), \quad i = 1,\ldots,n-2,$$

$$Je_{n-1} = \sum_{i=1}^{n-2} a^i e_i + \lambda e_n + normal\ part,$$

$$Je_n = \sum_{i=1}^{n-2} b^i e_i - \lambda e_{n-1} + normal\ part.$$

It follows immediately that $a^i = 0$, $b^i = 0$ for $i = 1,\ldots,n-2$ and that $\lambda = \bar{g}(Je_{n-1}, e_n)$.

Now we choose another orthonormal pair of vectors: $e'_{n-1}$ and $e'_n$. Then, for some $\theta$, we have

$$e'_{n-1} = e_{n-1}\cos\theta + e_n\sin\theta,$$

$$e'_n = -e_{n-1}\sin\theta + e_n\cos\theta$$

and consequently

$$\begin{aligned}
\lambda' &= \bar{g}(Je'_{n-1}, e'_n) \\
&= \bar{g}(Je_{n-1}\cos\theta + Je_n\sin\theta, -e_{n-1}\sin\theta + e_n\cos\theta) \\
&= \lambda\bar{g}(e_n\cos\theta - e_{n-1}\sin\theta, -e_{n-1}\sin\theta + e_n\cos\theta) \\
&= \lambda(\cos^2\theta + \sin^2\theta) = \lambda.
\end{aligned}$$

This shows that $\lambda$ is independent of the choice of $e_{n-1}$ and $e_n$.

Therefore, since $\lambda$ is not necessarily identically equal to zero, we conclude that a CR submanifold defined by Definition 7.2 is not always a CR submanifold defined in the sense of Definition 7.3. Especially, computing $\lambda$ in Example 7.6, we obtain

$$\lambda = \left\langle J\frac{\partial}{\partial x^1}, \frac{\partial}{\partial y^1} + \frac{\partial}{\partial x^2} \right\rangle = \left\langle \frac{\partial}{\partial y^1}, \frac{\partial}{\partial y^1} + \frac{\partial}{\partial x^2} \right\rangle = 1.$$

Since $\dim H(M) = 0$ and $\lambda$ is not identically equal to zero, we conclude that Example 7.6 is one of the examples of CR submanifolds in the sense of Definition 7.2, but not in the sense of Definition 7.3.

Now, let $M$ be a submanifold of a complex manifold. Then for any $X \in T(M)$, $J_i X$ can be written as a sum of the tangential part $iFX$ and the normal part $v(X)$ in the following way:

$$J_i X = iFX + v(X). \tag{7.3}$$

Then $F$ is an endomorphism on the tangent bundle $T(M)$ and $v$ is a normal bundle valued 1-form on $M$.

**Proposition 7.9.** *If $M$ is a CR submanifold in the sense of Definition 7.3, then the endomorphism $F$ satisfies $F^3 + F = 0$ and $\operatorname{rank} F = \dim \Delta$. Conversely, if the endomorphism $F$ satisfies $F^3 + F = 0$ and $\operatorname{rank} F$ is constant for $x \in M$, then $M$ is a CR submanifold in the sense of Definition 7.3.*

*Proof.* Let $M$ be a CR submanifold in the sense of Definition 7.3 and let $X \in T(M)$. Then we have $X = X_1 + X_2$, where $X_1 \in \Delta$ and $X_2 \in \Delta^\perp$. Using Definition 7.3, it follows that $v(X_1) = 0$ and $FX_2 = 0$ and consequently, using (7.3), we write

$$-\imath X_1 = J^2 \imath X_1 = \imath F^2 X_1 + v(FX_1), \qquad -\imath X_2 = J^2 \imath X_2 = Jv(X_2).$$

Separating and comparing the tangential parts and the normal parts, we conclude

$$F^2 X_1 = -X_1, \qquad v(FX_1) = 0.$$

Hence, $J\imath FX_1 = \imath F^2 X_1 \in T(M)$ and $FX_1 \in \Delta$. Therefore

$$J^2 \imath X = J\imath FX_1 + Jv(X_2) = \imath F^2 X_1 + v(FX_1) - \imath X_2 = \imath(F^2 X_1 - X_2),$$
$$J^3 \imath X = J\imath(F^2 X_1 - X_2) = \imath F(F^2 X_1 - X_2) + v(F^2 X_1 - X_2)$$
$$= \imath F^3 X_1 - v(X_2).$$

On the other hand, we have

$$F^3 X = F^3 X_1 + F^3 X_2 = F^3 X_1$$

and

$$v(X) = v(X_1) + v(X_2) = v(X_2),$$

which implies

$$J^3 \imath X = -J\imath X = -\imath FX - v(X) = \imath F^3 X - v(X).$$

Thus, $F$ satisfies $F^3 + F = 0$ and $\operatorname{rank} F = \dim \Delta$.

Conversely, suppose that $F$ satisfies $F^3 + F = 0$ and let rank of $F$ be $r$. We put

$$\Delta^\perp = \{X \in T(M) | FX = 0\},$$
$$\Delta = \{X \in T(M) | g(X, Y) = 0, \, Y \in \Delta^\perp\}.$$

Then, by definition, $J\Delta^\perp \subset T^\perp(M)$ and for $X \in \Delta$, $Y \in \Delta^\perp$ we obtain

$$\overline{g}(J\imath X, \imath Y) = -\overline{g}(\imath X, J\imath Y) = -\overline{g}(\imath X, v(Y)) = 0.$$

Hence $J\imath X \perp \Delta^\perp$ and therefore $J\Delta = \Delta$, and the proof is complete. □

Especially, let $M$ be a $J$-invariant submanifold of a Kähler manifold $(\overline{M}, J)$. That is, for any $X \in T(M)$, $J\imath X \in T(M)$ and $J$ induces the natural almost complex structure $J'$ on $M$ (see Example 7.2). First we note the following

**Lemma 7.1.** *If $M$ is a $J$-invariant submanifold of a complex manifold $(\overline{M}, J)$ with Hermitian metric $\overline{g}$, then for any $\xi \in T_x^\perp(M)$, it follows $J\xi \in T_x^\perp(M)$.*

*Proof.* Under the conditions stated above, it follows

$$0 = \overline{g}(\imath X, \xi) = \overline{g}(J\imath X, J\xi) = \overline{g}(\imath J'X, J\xi). \tag{7.4}$$

Since $J' : T_x(M) \rightarrow T_x(M)$ is an isomorphism, for any $Y \in T_x(M)$ there exists $X \in T_x(M)$ such that $Y = J'X$. Thus, equation (7.4) shows that for any $Y \in T_x(M)$, $\overline{g}(\imath Y, J\xi) = 0$ and therefore $J\xi \in T_x^\perp(M)$.     $\square$

**Theorem 7.2.** *A $J$-invariant submanifold $M$ of a Kähler manifold $(\overline{M}, J)$ is a Kähler manifold.*

*Proof.* Since $M$ is a $J$-invariant submanifold of a Kähler manifold $(\overline{M}, J)$, relation (7.2) and Theorem 4.2 imply

$$J\imath Y = \imath J'Y, \quad \overline{\nabla}_X J = 0, \tag{7.5}$$

for all $X, Y \in T(M)$. Therefore, differentiating covariantly the first relation and using the second relation of (7.5), the Gauss formula (5.1) implies

$$J\overline{\nabla}_X \imath Y = \imath \nabla_X(J'Y) + h(X, J'Y). \tag{7.6}$$

Using again the Gauss formula and the first relation of (7.5), we can rewrite relation (7.6) as

$$\imath J' \nabla_X Y + Jh(X, Y) = \imath(\nabla_X J')Y + \imath J' \nabla_X Y + h(X, J'Y). \tag{7.7}$$

Separating the tangential part and the normal part of relation (7.7), we conclude

$$Jh(X, Y) = h(X, J'Y) \tag{7.8}$$

and $\nabla_X J' = 0$. Therefore, $(M, J')$ is a Kähler manifold, by Theorem 4.2.     $\square$

*Example 7.7.* Let $M$ be a submanifold of a complex Euclidean space $\mathbf{C}^n$ defined by

$$M = \{(z_1, \ldots, z_m, z_{m+1}, \ldots, z_n) \in \mathbf{C}^n \mid z_{m+j} = f_j(z_1, \ldots, z_m), \tag{7.9}$$
$$j = 1, \ldots, n - m\} \quad \text{for} \quad m < n,$$

where $f_j$ are holomorphic functions on $\mathbf{C}^m$. Putting

$$z_i = x_i + \sqrt{-1}y_i, \quad i = 1, \ldots, m,$$
$$f_j = g_j + \sqrt{-1}h_j, \quad j = 1, \ldots, n - m,$$

we can identify $M$ with a submanifold of $\mathbf{E}^{2n}$ defined by

$$M = \{(x_1, y_1, \ldots, x_n, y_n) \in \mathbf{E}^{2n} | x_{m+j} = g_j(x_1, y_1, \ldots, x_m, y_m),$$
$$y_{m+j} = h_j(x_1, y_1, \ldots, x_m, y_m)\}, j = 1, \ldots, n - m. \tag{7.10}$$

Let $(u_1, v_1, \ldots, u_m, v_m)$ be a local coordinate system of $M$. Then the defining equation for $M$ is given by

$$x_i = u_i, \quad y_i = v_i, \quad i = i, \ldots, m,$$
$$x_{m+j} = g_j(x_1, y_1, \ldots, x_m, y_m),$$
$$y_{m+j} = h_j(x_1, y_1, \ldots, x_m, y_m), \quad j = 1, \ldots, n - m,$$

and consequently

$$\frac{\partial x_k}{\partial u_i} = \frac{\partial y_k}{\partial v_i} = \delta_i^k, \quad \frac{\partial x_k}{\partial v_i} = \frac{\partial y_k}{\partial u_i} = 0, \tag{7.11}$$

$$\frac{\partial x_{m+j}}{\partial u_i} = \frac{\partial g_j}{\partial u_i}, \quad \frac{\partial x_{m+j}}{\partial v_i} = \frac{\partial g_j}{\partial v_i}, \tag{7.12}$$

$$\frac{\partial y_{m+j}}{\partial u_i} = \frac{\partial h_j}{\partial u_i}, \quad \frac{\partial y_{m+j}}{\partial v_i} = \frac{\partial h_j}{\partial v_i}, \tag{7.13}$$

for $k = 1, \ldots, m, i = i, \ldots, m, j = 1, \ldots, n - m$.

The tangent space $T_x(M)$ at $x \in M$ is spanned by

$$\left\{ \left( \frac{\partial}{\partial u_i} \right)_x, \left( \frac{\partial}{\partial v_i} \right)_x, i = 1, \ldots, m \right\}. \tag{7.14}$$

For the immersion $\imath : M \to \mathbf{E}^{2n}$, we have

$$\imath \left( \frac{\partial}{\partial u_i} \right) = \sum_{\lambda=1}^{n} \left( \frac{\partial x_\lambda}{\partial u_i} \frac{\partial}{\partial x_\lambda} + \frac{\partial y_\lambda}{\partial u_i} \frac{\partial}{\partial y_\lambda} \right)$$

$$= \sum_{k=1}^{m} \left( \frac{\partial x_k}{\partial u_i} \frac{\partial}{\partial x_k} + \frac{\partial y_k}{\partial u_i} \frac{\partial}{\partial y_k} \right) + \sum_{j=1}^{n-m} \left( \frac{\partial g_j}{\partial u_i} \frac{\partial}{\partial x_{m+j}} + \frac{\partial h_j}{\partial u_i} \frac{\partial}{\partial y_{m+j}} \right).$$

Therefore, using (7.11)–(7.13), we compute

$$\imath \left( \frac{\partial}{\partial u_i} \right) = \frac{\partial}{\partial x_i} + \sum_{j=1}^{n-m} \left( \frac{\partial g_j}{\partial u_i} \frac{\partial}{\partial x_{m+j}} + \frac{\partial h_j}{\partial u_i} \frac{\partial}{\partial y_{m+j}} \right) \tag{7.15}$$

and similarly

$$\imath \left( \frac{\partial}{\partial v_i} \right) = \frac{\partial}{\partial y_i} + \sum_{j=1}^{n-m} \left( \frac{\partial g_j}{\partial v_i} \frac{\partial}{\partial x_{m+j}} + \frac{\partial h_j}{\partial v_i} \frac{\partial}{\partial y_{m+j}} \right). \tag{7.16}$$

Applying $J$ to (7.15), we have

$$J\imath\left(\frac{\partial}{\partial u_i}\right) = \frac{\partial}{\partial y_i} + \sum_{j=1}^{n-m}\left(\frac{\partial g_j}{\partial u_i}\frac{\partial}{\partial y_{m+j}} - \frac{\partial h_j}{\partial u_i}\frac{\partial}{\partial x_{m+j}}\right). \tag{7.17}$$

Since $f_j$ are holomorphic functions, they satisfy the Cauchy-Riemann equations, namely,

$$\frac{\partial g_j}{\partial u_i} = \frac{\partial h_j}{\partial v_i}, \quad \frac{\partial g_j}{\partial v_i} = -\frac{\partial h_j}{\partial u_i},$$

and using (7.16) and (7.17), we compute

$$J\imath\left(\frac{\partial}{\partial u_i}\right) = \frac{\partial}{\partial y_i} + \sum_{j=1}^{n-m}\left(\frac{\partial g_j}{\partial v_i}\frac{\partial}{\partial x_{m+j}} + \frac{\partial h_j}{\partial v_i}\frac{\partial}{\partial y_{m+j}}\right) = \imath\left(\frac{\partial}{\partial v_i}\right). \tag{7.18}$$

In entirely the same way, we obtain

$$J\imath\left(\frac{\partial}{\partial v_i}\right) = -\imath\left(\frac{\partial}{\partial u_i}\right). \tag{7.19}$$

Using (7.18) and (7.19) we conclude that $JT(M) \subset T(M)$. Hence we deduce that $M$ is a $J$-invariant submanifold of $\mathbf{C}^n$. Since $\mathbf{C}^n$ is a Kähler manifold, using Theorem 7.2, it follows that $M$ is a Kähler manifold. $\diamond$

According to Theorem 7.2, we can choose an orthonormal normal basis $\xi_a$, $\xi_{a^*} \in T_x^\perp(M)$, in such a way that $\xi_{a^*} = J\xi_a$, for $a = 1,\ldots,q$, where $q = \frac{p}{2}$. Consequently, we can write

$$h(X,Y) = \sum_{a=1}^{q}\{g(A_aX,Y)\xi_a + g(A_{a^*}X,Y)\xi_{a^*}\} \tag{7.20}$$

where $A_{a^*}$ denotes the shape operator with respect to $\xi_{a^*}$. Thus, combining (7.8) with (7.20) gives

$$Jh(X,Y) = \sum_{a=1}^{q}\{g(A_aX,Y)\xi_{a^*} - g(A_{a^*}X,Y)\xi_a\}$$

$$= \sum_{a=1}^{q}\{g(A_aX,J'Y)\xi_a + g(A_{a^*}X,J'Y)\xi_{a^*}\}$$

$$= h(X,J'Y),$$

and therefore

$$g(A_aX,J'Y) = -g(A_{a^*}X,Y), \quad g(A_aX,Y) = g(A_{a^*}X,J'Y),$$

that is, $A_{a^*} = J'A_a$, $A_a = -J'A_{a^*}$. Since $J'$ is skew-symmetric and $A_a$ and $A_{a^*}$ are symmetric, we conclude

$$\text{trace }A_a = -\text{trace }J'A_{a^*} = 0, \quad \text{trace }A_{a^*} = \text{trace }J'A_a = 0 \quad \text{for} \quad a = 1,\ldots,q.$$

Thus, using Proposition 5.4, we have proved

**Theorem 7.3.** *A J-invariant submanifold of a Kähler manifold is a minimal submanifold.*

Here we note that

$$g(A_a J' X, Y) = -g(X, J' A_a Y) = -g(X, A_{a^*} Y)$$
$$= -g(A_{a^*} X, Y) = -g(J' A_a X, Y)$$

and an analogous consideration implies

$$A_a J' = -J' A_a, \quad A_{a^*} J' = -J' A_{a^*}. \tag{7.21}$$

# 8

# The Levi form

Let $\overline{M}$ be a complex manifold with the natural almost complex structure $J$ and let $M$ be a real submanifold of $\overline{M}$. In this section we consider the involutivity of the complexification of a holomorphic tangent bundle. For this purpose, the Levi form plays a very important role.

Let $H_x^C(M)$ be the complexification of the holomorphic tangent space $H_x(M)$ and

$$H_x^{(0,1)}(M) = \left\{ \imath X + \sqrt{-1}\, J\imath X \,|\, X \in H_x(M) \right\},$$
$$H_x^{(1,0)}(M) = \left\{ \imath X - \sqrt{-1}\, J\imath X \,|\, X \in H_x(M) \right\}.$$

Then we have

$$H_x^C(M) = H_x^{(0,1)}(M) \oplus H_x^{(1,0)}(M).$$

We define the following subbundles of the complexification of the tangent bundle $T^C(M)$

$$H^C(M) = \bigcup_{x \in M} H_x^C(M), \quad H^{(0,1)}(M) = \bigcup_{x \in M} H_x^{(0,1)}(M),$$
$$H^{(1,0)}(M) = \bigcup_{x \in M} H_x^{(1,0)}(M).$$

We begin with a well-known result.

**Proposition 8.1.** *Both distributions* $H^{(0,1)}(M)$ *and* $H^{(1,0)}(M)$ *are involutive.*

*Proof.* We only show that $H^{(0,1)}(M)$ is involutive, because the other case can be proved in entirely the same way. We compute $[V, W]$ for $V, W \in H^{(0,1)}(M)$. Then, for some $X, Y \in H(M)$, we have

$$[V, W] = [\imath X + \sqrt{-1}\, J\imath X, \imath Y + \sqrt{-1}\, J\imath Y]$$
$$= [\imath X, \imath Y] - [J\imath X, J\imath Y] + \sqrt{-1}([J\imath X, \imath Y] + [\imath X, J\imath Y]). \quad (8.1)$$

M. Djorić, M. Okumura, *CR Submanifolds of Complex Projective Space*,      53
Developments in Mathematics 19, DOI 10.1007/978-1-4419-0434-8_8,
© Springer Science+Business Media, LLC 2010

Since $\overline{M}$ is a complex manifold, the Nijenhuis tensor

$$\overline{N}(\imath X, \imath Y) = J[\imath X, \imath Y] - J[J\imath X, J\imath Y] - [\imath X, J\imath Y] - [J\imath X, \imath Y]$$

vanishes identically and we obtain

$$[\imath X, \imath Y] - [J\imath X, J\imath Y] = J(-[\imath X, J\imath Y] - [J\imath X, \imath Y]),$$
$$[\imath X, J\imath Y] + [J\imath X, \imath Y] = J([\imath X, \imath Y] - [J\imath X, J\imath Y]). \tag{8.2}$$

Now, using (8.1) and (8.2), we get

$$[V, W] = [\imath X, \imath Y] - [J\imath X, J\imath Y] + \sqrt{-1}\, J([\imath X, \imath Y] - [J\imath X, J\imath Y]).$$

Since $X, Y \in H(M)$, it follows that $J\imath X, J\imath Y \in T(M)$ and therefore we conclude

$$[\imath X, \imath Y] - [J\imath X, J\imath Y] \in T(M).$$

Also, from the above discussions, it follows

$$[\imath X, \imath Y] - [J\imath X, J\imath Y] \in JT(M)$$

and this implies that $H^{(0,1)}(M)$ is involutive. $\qquad\square$

**Proposition 8.2.** *If $H(M)$ is involutive, the integral submanifold of $H(M)$ is a complex manifold.*

*Proof.* If $X \in H(M)$, then $J\imath X = \imath F X + v(X) = \imath F X$ and $F^2 = -id$. Thus $F$ is an almost complex structure on the integral submanifold. Since the ambient manifold $\overline{M}$ is a complex manifold, the Nijenhuis tensor $\overline{N}(\imath X, \imath Y) = 0$. Therefore we have

$$[\imath X, \imath Y] - [J\imath X, J\imath Y] + J[J\imath X, \imath Y] + J[\imath X, J\imath Y]$$
$$= \imath[X, Y] - \imath[FX, FY] + J\imath[FX, Y] + J\imath[X, FY] = 0,$$

that is,

$$\imath[X, Y] - \imath[FX, FY] + \imath F[FX, Y] + v([FX, Y]) + \imath F[X, FY] + v([X, FY]) = 0.$$

The tangential part of the last equation is just the Nijenhuis tensor $N(X, Y)$. Hence the integral submanifold is a complex manifold with almost complex structure $F$. $\qquad\square$

**Lemma 8.1.** *The normal part of $J[J\imath X, \imath Y]$ is equal to the normal part of $J[J\imath Y, \imath X]$.*

*Proof.* From relation (8.2), we have

$$\imath[X, Y] - \imath[FX, FY] = J([J\imath Y, \imath X] - [J\imath X, \imath Y]).$$

The left hand members of the last equation are tangent to $M$ and therefore the normal part of the right hand members must be zero. This completes the proof. $\qquad\square$

As we have shown, both $H^{(0,1)}(M)$ and $H^{(1,0)}(M)$ are involutive. But this does not imply that $H^C(M) = H^{(0,1)}(M) \oplus H^{(1,0)}(M)$ is involutive and therefore let us consider the involutivity of $H^C(M)$.

Let $V \in H^{(0,1)}(M)$, $W \in H^{(1,0)}(M)$. Then, for some $X, Y \in H(M)$, $V = \imath X + \sqrt{-1}\, J\imath X$ and $W = \imath Y - \sqrt{-1}\, J\imath Y$ and consequently

$$[V, W] = [\imath X, \imath Y] + [J\imath X, J\imath Y] - \sqrt{-1}([\imath X, J\imath Y] - [J\imath X, \imath Y]).$$

Since $X, Y \in H(M)$, it follows $J\imath X = \imath F X$, $J\imath Y = \imath F Y$ and therefore $[J\imath X, J\imath Y]$, $[\imath X, J\imath Y]$, $[J\imath X, \imath Y] \in T(M)$. Hence $[V, W] \in T^C(M)$. However, in general, $[V, W] \notin JT^C(M)$.

**Lemma 8.2.** *Under the above assumptions, a necessary and sufficient condition for* $[V, W] \in H^C(M)$ *is that* $J[\imath X, J\imath Y] - J[J\imath X, \imath Y] \in T(M)$.

*Proof.* First we note that $J[\imath X, J\imath Y] - J[J\imath X, \imath Y] \in T(M)$ is equivalent to $[\imath X, J\imath Y] - [J\imath X, \imath Y] \in JT(M)$. Then by definition of $H^C(M)$, the necessity is trivial. To prove the sufficiency, we take $X' \in H(M)$. $X' \in JT(M)$ implies that there exists $X \in T(M)$ such that $\imath X' = J\imath X$ and we have

$$[\imath X', \imath Y] + [J\imath X', J\imath Y] = [J\imath X, \imath Y] + [J^2\imath X, J\imath Y]$$
$$= [J\imath X, \imath Y] - [\imath X, J\imath Y] \in JT(M).$$

This completes the proof.    □

The Levi form is defined in such a way that it measures the degree to which $H^C(M)$ fails to be involutive. As we have shown, the normal part of $J[J\imath X, \imath Y]$ is equal to that of $J[J\imath Y, \imath X]$. We give the following definition of Levi form.

**Definition 8.1.** The *Levi form* $L$ is the projection of $J[J\imath X, \imath Y]$ to $T^\perp(M)$ for $X, Y \in H(M)$.

**Theorem 8.1.** *[31] Let $\overline{M}$ be a complex manifold with torsion-free affine connection $\overline{\nabla}$ whose parallel translation leaves the almost complex structure $J$ invariant and $M$ be a real submanifold of $\overline{M}$. Then we have*

$$L(X, Y) = h(X, Y) + h(FX, FY), \tag{8.3}$$

*for $X, Y \in H(M)$, where $h$ denotes the second fundamental form with respect to $\overline{\nabla}$ and $F$ is defined by (7.3).*

*Proof.* As we have shown, $X \in H(M)$, implies $J\imath X = \imath F X$ and therefore, using the Gauss formula (5.1), we compute

$$
\begin{aligned}
J[J\imath X, \imath Y] = J[\imath FX, \imath Y] &= J\imath[FX, Y] \\
&= J\imath\left(\nabla_{FX}Y - \nabla_Y(FX)\right) \\
&= J\left(\overline{\nabla}_{FX}\imath Y - h(FX, Y) - \overline{\nabla}_{Y}\imath FX + h(Y, FX)\right) \\
&= J\left(\overline{\nabla}_{FX}\imath Y - \overline{\nabla}_{Y}\imath FX\right) \\
&= \overline{\nabla}_{FX}(J\imath Y) - \overline{\nabla}_Y(J\imath FX) \\
&= \overline{\nabla}_{FX}\imath FY - \overline{\nabla}_{Y}\imath F^2 X \\
&= \overline{\nabla}_{FX}\imath FY + \overline{\nabla}_{Y}\imath X \\
&= \imath\nabla_{FX}(FY) + h(FX, FY) + \imath\nabla_Y X + h(Y, X).
\end{aligned}
$$

Using Definition 8.1, this establishes the formula (8.3).    □

# The principal circle bundle $S^{2n+1}(P^n(C), S^1)$

It is well known that an odd-dimensional sphere is a circle bundle over the complex projective space (see [33]). Consequently, many geometric properties of the complex projective space are inherited from those of the sphere. Especially, at the end of this section, we prove that the complex projective space has constant holomorphic sectional curvature.

Let $C^{n+1}$ be the $(n+1)$-dimensional complex space with natural Kähler structure $(J', \langle, \rangle)$ recalled in Example 4.2 and let $S^{2n+1}$ be the *unit sphere* defined by

$$S^{2n+1} = \{(z^1, \ldots, z^{n+1}) \in C^{n+1} \mid \sum_{i=1}^{n+1} z^i \bar{z}^i = 1\}$$

$$= \{(x^1, y^1, \ldots, x^{n+1}, y^{n+1}) \in R^{2n+2} \mid \sum_{i=1}^{n+1} [(x^i)^2 + (y^i)^2] = 1\}.$$

The unit normal vector field $\xi$ to $S^{2n+1}$ is given by

$$\xi = -\sum_{i=1}^{n+1} \left( x^i \frac{\partial}{\partial x^i} + y^i \frac{\partial}{\partial y^i} \right).$$

From

$$\langle J'\xi, \xi \rangle = \langle J'^2\xi, J'\xi \rangle = -\langle \xi, J'\xi \rangle,$$

it follows $\langle J'\xi, \xi \rangle = 0$, that is, $J'\xi \in T(S^{2n+1})$. We put

$$J'\xi = -\imath V', \tag{9.1}$$

where $\imath$ denotes the immersion of $S^{2n+1}$ into $C^{n+1}$. From the Hermitian property of $\langle, \rangle$, it is easily seen that $V'$ is a unit tangent vector field of $S^{2n+1}$ and with respect to the natural basis, $\imath V'$ is represented by

$$\imath V' = \sum_{i=1}^{n+1} \left( -y^i \frac{\partial}{\partial x^i} + x^i \frac{\partial}{\partial y^i} \right).$$

M. Djorić, M. Okumura, *CR Submanifolds of Complex Projective Space*, Developments in Mathematics 19, DOI 10.1007/978-1-4419-0434-8_9, © Springer Science+Business Media, LLC 2010

Let us denote by $g'$ the metric on $\mathbf{S}^{2n+1}$, induced by the metric $\langle,\rangle$. Defining the 1-form $u'$ on $\mathbf{S}^{2n+1}$ by

$$u'(X') = g'(V', X') = \langle \imath V', \imath X'\rangle, \quad \text{for} \quad X' \in T(S^{2n+1}),$$

we can write

$$u' = \sum_{i=1}^{n+1}(-y^i dx^i + x^i dy^i).$$

Let $v^i$ be the $i$-th component of $\imath V'$ with respect to complex coordinates $z^i = x^i + \sqrt{-1}y^i$ of $\mathbf{C}^{n+1}$. Then, $\imath V'$ is represented as a position vector field by $v^i = \sqrt{-1}\,z^i$ and consequently, the integral curve of $\imath V'$ is a great circle

$$\mathbf{S}^1 = \{e^{\sqrt{-1}\theta}\,|\,0 \le \theta < 2\pi\}.$$

We define a map $\mathbf{S}^{2n+1} \times \mathbf{S}^1 \to \mathbf{S}^{2n+1}$ by

$$(z, e^{\sqrt{-1}\theta}) \mapsto ze^{\sqrt{-1}\theta}.$$

Then, $\mathbf{S}^1$ acts on $\mathbf{S}^{2n+1}$ freely and the quotient space of $\mathbf{S}^{2n+1}$ by the equivalence relation induced by $\mathbf{S}^1$ is the complex projective space $\mathbf{P}^n(\mathbf{C})$. Thus we get the *principal circle bundle* $\mathbf{S}^{2n+1}(\mathbf{P}^n(\mathbf{C}), \mathbf{S}^1)$. We put

$$H_p(S^{2n+1}) = \{X' \in T_p(\mathbf{S}^{2n+1})\,|\,u'(X') = 0\}.$$

Then $u'$ defines a connection form of the principal circle bundle $\mathbf{S}^{2n+1}(\mathbf{P}^n(\mathbf{C}), \mathbf{S}^1)$ and we have

$$T_p(\mathbf{S}^{2n+1}) = H_p(\mathbf{S}^{2n+1}) \oplus \text{span}\{V'_p\}.$$

We call $H_p(\mathbf{S}^{2n+1})$ and $\text{span}\{V'_p\}$ the horizontal subspace and the vertical subspace of $T_p(\mathbf{S}^{2n+1})$, respectively. By definition, the horizontal subspace $H_p(\mathbf{S}^{2n+1})$ is isomorphic to $T_{\pi(p)}(\mathbf{P}^n(\mathbf{C}))$, where $\pi$ is the natural projection from $\mathbf{S}^{2n+1}$ onto $\mathbf{P}^n(\mathbf{C})$. Therefore, for a vector field $X$ on $\mathbf{P}^n(\mathbf{C})$, there exists unique horizontal vector field $X'$ of $\mathbf{S}^{2n+1}$ such that $\pi(X') = X$. The vector field $X'$ is called the *horizontal lift* of $X$ and we denote it by $X^*$.

**Proposition 9.1.** *As a subspace of $T_p(\mathbf{C}^{n+1})$, $H_p(\mathbf{S}^{2n+1})$ is a $J'$-invariant subspace.*

*Proof.* By definition (9.1) of the vertical vector field $V'$, for $X' \in H_p(\mathbf{S}^{2n+1})$, it follows

$$\langle J'\imath X', \xi \rangle = -\langle \imath X', J'\xi \rangle = \langle \imath X', \imath V' \rangle = 0.$$

This shows that $J'\imath X' \in T_p(\mathbf{S}^{2n+1})$. In entirely the same way we compute

$$\langle J'\imath X', \imath V' \rangle = -\langle \imath X', J'\imath V' \rangle = \langle \imath X', -\xi \rangle = 0$$

and hence $J'\imath X' \in H(\mathbf{S}^{2n+1})$, which completes the proof. $\qquad\square$

Therefore, the almost complex structure $J$ can be induced on $T_{\pi(p)}\mathbf{P}^n(\mathbf{C})$ and we set

$$(JX)^* = J'\imath X^*. \tag{9.2}$$

Next, using the Gauss formula (5.1) for the vertical vector field $V'$ and a horizontal vector field $X'$ of $T_p(\mathbf{S}^{2n+1})$, we compute

$$\begin{aligned}
\nabla^E_{X'}\imath V' &= \imath\nabla'_{X'}V' + g'(A'X', V')\xi \\
&= \imath\nabla'_{X'}V' + \langle \imath X', \imath V'\rangle\xi \\
&= \imath\nabla'_{X'}V',
\end{aligned} \tag{9.3}$$

where $\nabla^E$ denotes the Euclidean connection of $\mathbf{E}^{2n+2}$, $\nabla'$ denotes the connection of $\mathbf{S}^{2n+1}$ and $A'$ denotes the shape operator with respect to $\xi$. Now, using relations (9.3), (9.1) and the Weingarten formula (5.6), we conclude

$$\begin{aligned}
\nabla'_{X'}V' &= -\nabla^E_{X'}(J'\xi) = -J'\nabla^E_{X'}\xi \\
&= J'(\imath A'X') = J'\imath X'.
\end{aligned} \tag{9.4}$$

Consequently, according to notation (9.2), relation (9.4) can be written as

$$\nabla'_{X*}V' = (JX)^*. \tag{9.5}$$

We note that, since by definition, the Lie derivative of a horizontal lift of a vector field with respect to a vertical vector field is zero, it follows

$$0 = L_{V'}X^* = [V', X^*] = \nabla'_{V'}X^* - \nabla'_{X*}V'$$

and using (9.5), we conclude

$$\nabla'_{V'}X^* = (JX)^*. \tag{9.6}$$

We define a Riemannian metric $g$ and a connection $\nabla$ in $\mathbf{P}^n(\mathbf{C})$ respectively by

$$g(X, Y) = g'(X^*, Y^*), \tag{9.7}$$
$$\nabla_X Y = \pi(\nabla'_{X*}Y^*). \tag{9.8}$$

Then $(\nabla_X Y)^*$ is the horizontal part of $\nabla'_{X*}Y^*$ and therefore

$$\nabla'_{X*}Y^* = (\nabla_X Y)^* + g'(\nabla'_{X*}Y^*, V')V'. \tag{9.9}$$

Using relations (9.5) and (9.7), we compute

$$g'(\nabla'_{X*}Y^*, V') = -g'(Y^*, \nabla'_{X*}V') = -g'(Y^*, (JX)^*) = -g(Y, JX),$$

and, using (9.9), we conclude

$$\nabla'_{X*}Y^* = (\nabla_X Y)^* - g(JX, Y)V'. \tag{9.10}$$

**Proposition 9.2.** $\nabla$ *is the Levi-Civita connection for* $g$.

*Proof.* Let $T$ be the torsion tensor field of $\nabla$. Then we have

$$T(X,Y) = \nabla_X Y - \nabla_Y X - [X,Y] = \pi \nabla'_{X*} Y^* - \pi \nabla'_{Y*} X^* - [\pi X^*, \pi Y^*]$$
$$= \pi(\nabla'_{X*} Y^* - \nabla'_{Y*} X^* - [X^*, Y^*]) = \pi(T'(X^*, Y^*)) = 0.$$

Hence $\nabla$ is torsion-free. We now show that $\nabla$ is a metric connection.

$$(\nabla_X g)(Y,Z) = X(g(Y,Z)) - g(\nabla_X Y, Z) - g(Y, \nabla_X Z)$$
$$= X^*(g'(Y^*, Z^*)) - g'((\nabla_X Y)^*, Z^*) - g'(Y^*, (\nabla_X Z)^*).$$

Since $Z^*$ is horizontal, using relation (9.9), it follows

$$g'((\nabla_X Y)^*, Z^*) = g'(\nabla'_{X*} Y^*, Z^*)$$

and we compute

$$(\nabla_X g)(X,Y) = X^*(g'(Y^*, Z^*)) - g'(\nabla'_{X*} Y^*, Z^*) - g'(Y^*, \nabla'_{X*} Z^*)$$
$$= (\nabla'_{X*} g')(Y^*, Z^*) = 0,$$

where we have used the fact that $\nabla'$ is the Levi-Civita connection for $g'$. Thus, $\nabla$ is the Levi-Civita connection for $g$ and the proof is complete. $\qquad\square$

Further, let $\Gamma$ be a curve on $\mathbf{S}^{2n+1}$ whose tangent vector field $\frac{d\Gamma}{ds}$ is horizontal and put

$$\gamma(s) = \pi(\Gamma(s)).$$

Then $\gamma(s)$ is a curve in $\mathbf{P}^n(\mathbf{C})$ and the tangent vector field $\dot\gamma$ of $\gamma(s)$ is $\pi(\frac{d\Gamma}{ds})$. Therefore $\frac{d\Gamma}{ds}$ is the horizontal lift of $\dot\gamma$ and

$$\nabla_{\dot\gamma} \dot\gamma = \pi\left(\nabla'_{\frac{d\Gamma}{ds}} \frac{d\Gamma}{ds}\right).$$

Hence, if $\Gamma(s)$ is a geodesic of $\mathbf{S}^{2n+1}$, then $\gamma(s)$ is a geodesic of $\mathbf{P}^n(\mathbf{C})$.

Conversely, let $\gamma(s)$ be a geodesic of $\mathbf{P}^n(\mathbf{C})$ through a point $x \in \mathbf{P}^n(\mathbf{C})$. Then through any point $w \in \pi^{-1}(x) \subset \mathbf{S}^{2n+1}$, there exists a unique geodesic $\Gamma(s)$ whose tangent vector at $w$ is $\dot\gamma^*(0)$. Thus, $\Gamma(s)$ is the horizontal lift of the geodesic $\gamma(s)$ and it may be expressed as

$$\Gamma(s) = w \cos s + \dot\gamma^* \sin s,$$

where we regard that $w$ is the position vector at the initial point $w \in \mathbf{S}^{2n+1} \subset \mathbf{C}^{n+1} = \mathbf{E}^{2n+2}$. Hence any geodesic $\gamma(s)$ of $\mathbf{P}^n(\mathbf{C})$ is written as

$$\gamma(s) = \pi(w \cos s + \dot\gamma^* \sin s).$$

By virtue of (9.10), it follows

$$
\begin{aligned}
[X^*, Y^*] &= [X, Y]^* + g'([X^*, Y^*], V')V' \\
&= [X, Y]^* + g'(\nabla'_{X^*} Y^* - \nabla'_{Y^*} X^*, V')V' \\
&= [X, Y]^* + g'((\nabla_X Y)^* - g(JX, Y)V' - (\nabla_Y X)^* \\
&\quad + g(JY, X)V', V')V'
\end{aligned}
$$

and therefore

$$
[X^*, Y^*] = [X, Y]^* - 2g(JX, Y)V'. \tag{9.11}
$$

Consequently, using (9.5), (9.6), (9.10) and (9.11), the curvature tensor $R$ of $\mathbf{P}^n(\mathbf{C})$ is calculated as follows:

$$
\begin{aligned}
R(X, Y)Z &= \nabla_X \nabla_Y Z - \nabla_Y \nabla_X Z - \nabla_{[X,Y]} Z \\
&= \pi\{\nabla'_{X^*}(\nabla_Y Z)^* - \nabla'_{Y^*}(\nabla_X Z)^* - \nabla'_{[X,Y]^*} Z^*\} \\
&= \pi\{\nabla'_{X^*}(\nabla'_{Y^*} Z^* + g(JY, Z)V') - \nabla'_{Y^*}(\nabla'_{X^*} Z^* + g(JX, Z)V') \\
&\quad - \nabla'_{[X^*, Y^*] + 2g(JX,Y)V'} Z^*\} \\
&= \pi\{\nabla'_{X^*} \nabla'_{Y^*} Z^* + g(JY, Z)\nabla'_{X^*} V' - \nabla'_{Y^*} \nabla'_{X^*} Z^* \\
&\quad - g(JX, Z)\nabla'_{Y^*} V' - \nabla'_{[X^*, Y^*]} Z^* - 2g(JX, Y)\nabla'_{V'} Z^*\} \\
&= \pi\{R'(X^*, Y^*)Z^* + g(JY, Z)J'\imath X^* - g(JX, Z)J'\imath Y^* \\
&\quad - 2g(JX, Y)J'\imath Z^*\}.
\end{aligned}
$$

Since the curvature tensor $R'$ of $\mathbf{S}^{2n+1}$ satisfies

$$
\begin{aligned}
R'(X^*, Y^*)Z^* &= g'(Y^*, Z^*)X^* - g'(X^*, Z^*)Y^* \\
&= g(Y, Z)X^* - g(X, Z)Y^*, \tag{9.12}
\end{aligned}
$$

we conclude that the curvature tensor of $\mathbf{P}^n(\mathbf{C})$ is given by

$$
\begin{aligned}
R(X, Y)Z &= g(Y, Z)X - g(X, Z)Y + g(JY, Z)JX \\
&\quad - g(JX, Z)JY - 2g(JX, Y)JZ. \tag{9.13}
\end{aligned}
$$

Let $K_{XY}$ be the sectional curvature of $\mathbf{P}^n(\mathbf{C})$. Then, by definition of the *sectional curvature* and (9.13), it follows that

$$
\begin{aligned}
K_{XY} &= \frac{g(R(X, Y)Y, X)}{g(X, X)g(Y, Y) - g(X, Y)^2} \\
&= \frac{g(Y, Y)g(X, X) - g(X, Y)^2 + 3g(JX, Y)^2}{g(X, X)g(Y, Y) - g(X, Y)^2} \\
&= 1 + \frac{3g(JX, Y)^2}{g(X, X)g(Y, Y) - g(X, Y)^2} = 1 + 3\cos\theta,
\end{aligned}
$$

where $\theta$ is the angle between the planes $\{X, Y\}$ and $\{JX, JY\}$. Since $0 \leq \theta \leq \frac{\pi}{2}$, in $\mathbf{P}^n(\mathbf{C})$, we conclude

$$
1 \leq K_{XY} \leq 4. \tag{9.14}
$$

The *holomorphic sectional curvature* $H(X)$ of a complex manifold is defined by

$$H(X) = K_{X,JX} = \frac{g(R(X,JX)JX,X)}{g(X,X)^2}. \tag{9.15}$$

The holomorphic sectional curvature of $\mathbf{P}^n(\mathbf{C})$ is 4 since

$$H(X) = 1 + \frac{3g(X,X)^2}{g(X,X)^2} = 4. \tag{9.16}$$

More generally, a Kähler manifold $M$ is called a *complex space form* if it has constant holomorphic sectional curvature, namely, if $H(X)$ is a constant for all $J$-invariant planes $\{X, JX\}$ in $T_x(M)$ and for all points $x \in M$.

Now, let the holomorphic sectional curvature $H(X)$ be independent of $X \in T_x(M)$ for all $x \in M$, namely, let $k(x)$ be a real-valued function on $M$ such that $H(X) = 4\,k(x)$. Then by (9.15) we have

$$g(R(X,JX)JX,X) = 4\,k(x)g(X,X)^2,$$

for any $X \in T_x(M)$, from which it follows

$$\sum_P g(R(X,JY)JZ,W) = 4\,k(x) \sum_P g(X,Y)g(Z,W) \tag{9.17}$$

where $\sum_P$ denotes the sum of all permutations with respect to $X,Y,Z,W \in T_x(M)$.

Since $M$ is a Kähler manifold, Theorem 4.2 implies $\nabla_X J = 0$ for any $X$ and therefore,

$$R(X,Y)JZ = JR(X,Y)Z.$$

Hence, we compute

$$g(R(X,Y)JZ,W) = g(JR(X,Y)Z,W) = -g(R(X,Y)Z,JW)$$
$$= g(R(X,Y)JW,Z). \tag{9.18}$$

Now, relation (9.17) and repeated application of the property

$$g(R(X,Y)Z,W) = g(R(W,Z)Y,X)$$

of the curvature tensor $R$ and relation (9.18) imply

$$g(R(X,JY)JW,Z) + g(R(X,JZ)JY,W) + g(R(X,JW)JY,Z)$$
$$= 4\,k(x)\{g(X,Y)g(Z,W) + g(X,Z)g(Y,W) + g(X,W)g(Y,Z)\}.$$

Substituting $JY$ and $JW$ for $Y$ and $W$ in the above equation, respectively, we get

$$g(R(X,Y)W,Z) - g(R(X,JZ)Y,JW) + g(R(X,W)Y,Z)$$
$$= 4\,k(x)\{g(X,JY)g(Z,JW) + g(X,Z)g(Y,W) + g(X,JW)g(JY,Z)\}.$$

Taking the skew symmetric part of this equation with respect to $Z$ and $W$ and using the Bianchi identity, we obtain

$$2g(R(X,Y)W,Z) + g(R(X,JZ)JW,Y) - g(R(X,JW)JZ,Y)$$
$$+g(R(X,W)Y,Z) + g(R(X,Z)W,Y)$$
$$= 4\,k(x)\{-2g(X,JY)g(JZ,W) - g(X,W)g(Y,Z) \qquad (9.19)$$
$$+g(X,Z)g(Y,W) - g(X,JZ)g(JY,W) + g(X,JW)g(JY,Z)\}.$$

On the other hand, from (9.18) it follows

$$g(R(X,JZ)JW,Y) - g(R(X,JW)JZ,Y)$$
$$= g(R(X,JZ)JW,Y) + g(R(X,JW)Y,JZ)$$
$$= -g(R(X,Y)JZ,JW) = -g(JR(X,Y)Z,JW)$$
$$= -g(R(X,Y)Z,W). \qquad (9.20)$$

Using (9.20), relation (9.19) becomes

$$g(R(X,Y)Z,W) = k(x)\{g(X,W)g(Y,Z) - g(X,Z)g(Y,W)$$
$$+g(JY,Z)g(JX,W) - g(JX,Z)g(JY,W) - 2g(JX,Y)g(JZ,W)\},$$

that is, the Riemannian curvature tensor $R$ of a complex space form is given by

$$R(X,Y)Z = k(x)\,\{g(Y,Z)X - g(X,Z)Y + g(JY,Z)JX$$
$$- g(JX,Z)JY - 2g(JX,Y)JZ\}. \qquad (9.21)$$

Using the second Bianchi identity, it can be proved that $k(x)$ is constant.

Moreover, it is well-known that two complete, simply connected complex space forms of the same constant holomorphic sectional curvature are isometric and biholomorphic. Any Kähler manifold of constant holomorphic sectional curvature $k$ is locally isometric to one of the following spaces:

| | | | | |
|---|---|---|---|---|
| complex | Euclidean | space | $\mathbf{C}^n$, | $(k = 0)$, |
| complex | projective | space | $\mathbf{P}^n(\mathbf{C})$, | $(k > 0)$, |
| complex | hyperbolic | space | $\mathbf{H}^n(\mathbf{C})$, | $(k < 0)$. |

For the proof and more details we refer the reader to [33], [68].

# Submersion and immersion

As we have seen in Section 9, geometric properties of complex projective space are induced from that of an odd-dimensional sphere. Therefore, for studying submanifolds of a complex projective space, it is of great interest how to pull down some formulae deduced for submanifolds of a sphere to those for submanifolds of a complex projective space.

Let $M$ be an $n$-dimensional submanifold of $\mathbf{P}^{\frac{n+p}{2}}(\mathbf{C})$ and $\pi^{-1}(M)$ be the circle bundle over $M$ which is compatible with the Hopf map

$$\pi : \mathbf{S}^{n+p+1} \to \mathbf{P}^{\frac{n+p}{2}}(\mathbf{C}).$$

Then $\pi^{-1}(M)$ is a submanifold of $\mathbf{S}^{n+p+1}$. The compatibility with the Hopf map is expressed by $\pi \circ \imath' = \imath \circ \pi$, where $\imath$ and $\imath'$ are the immersions of $M$ into $\mathbf{P}^{\frac{n+p}{2}}(\mathbf{C})$ and $\pi^{-1}(M)$ into $\mathbf{S}^{n+p+1}$, respectively (see Section 9, but be aware that in Section 10 our notation of immersions is in conflict with that of Section 9).

Let $\xi_a$, $a = 1, \dots, p$ be orthonormal normal local fields to $M$ in $\mathbf{P}^{\frac{n+p}{2}}(\mathbf{C})$, and let $\xi_a^*$'s be the horizontal lifts of $\xi_a$. Then $\xi_a^*$'s are mutually orthonormal normal local fields to $\pi^{-1}(M)$ in $\mathbf{S}^{n+p+1}$. Consequently, using relation (9.7), at each point $y \in \pi^{-1}(M)$, we compute

$$g^S(\imath' X^*, \xi_a^*) = g^S((\imath X)^*, \xi_a^*) = \overline{g}(\imath X, \xi_a) = 0,$$
$$g^S(\imath' V, \xi_a^*) = g^S(V', \xi_a^*) = 0,$$
$$g^S(\xi_a^*, \xi_b^*) = \overline{g}(\xi_a, \xi_b) = \delta_{ab},$$

where $g^S$ and $\overline{g}$ denote the Riemannian metric on $\mathbf{S}^{n+p+1}$ and $\mathbf{P}^{\frac{n+p}{2}}(\mathbf{C})$, respectively. Here $V' = \imath'V$ is a unit tangent vector field of $\mathbf{S}^{n+p+1}$ defined by relation (9.1), namely, $J'\xi = -fV'$, where $f$ denotes the immersion of $\mathbf{S}^{n+p+1}$ into $\mathbf{C}^{\frac{n+p+2}{2}}$, $J'$ is the natural almost complex structure of $\mathbf{C}^{\frac{n+p+2}{2}}$ and $\xi$ is the unit normal vector field to $\mathbf{S}^{n+p+1}$.

M. Djorić, M. Okumura, *CR Submanifolds of Complex Projective Space*, Developments in Mathematics 19, DOI 10.1007/978-1-4419-0434-8_10, © Springer Science+Business Media, LLC 2010

Now, let $\nabla^S$, $\nabla'$, $\overline{\nabla}$ and $\nabla$ be the Riemannian connections of $\mathbf{S}^{n+p+1}$, $\pi^{-1}(M)$, $\mathbf{P}^{\frac{n+p}{2}}(\mathbf{C})$ and $M$, respectively. By means of the Gauss formula (5.1) and relations (7.3) and (9.10), we compute

$$\nabla^S_{X*}\imath'Y^* = \nabla^S_{X*}(\imath Y)^* = (\overline{\nabla}_X\imath Y)^* - \overline{g}(J\imath X, \imath Y)\imath'V$$
$$= (\imath\nabla_X Y + h(X,Y))^* + \overline{g}(\imath FX, \imath Y)\imath'V$$
$$= \imath'(\nabla_X Y)^* + (h(X,Y))^* + g(FX,Y)\imath'V, \qquad (10.1)$$

where $g$ is the metric on $M$. On the other hand, we also have

$$\nabla^S_{X*}\imath'Y^* = \imath'\nabla'_{X*}Y^* + h'(X^*, Y^*)$$
$$= \imath'((\nabla_X Y)^* + g(FX,Y)V) + h'(X^*, Y^*), \qquad (10.2)$$

where $h$ and $h'$ denote the second fundamental form of $M$ and $\pi^{-1}(M)$, respectively. Comparing the vertical part and the horizontal part of relations (10.1) and (10.2), we conclude

$$h'(X^*, Y^*) = (h(X,Y))^*, \qquad (10.3)$$

that is,

$$\sum_{a=1}^{p} g'(A'_a X^*, Y^*)\xi^*_a = \left(\sum_{a=1}^{p} g(A_a X, Y)\xi_a\right)^* = \sum_{a=1}^{p} g(A_a X, Y)\xi^*_a,$$

where $A_a$ and $A'_a$ are the shape operators with respect to normal vector fields $\xi_a$ and $\xi^*_a$ of $M$ and $\pi^{-1}(M)$, respectively. Consequently, we have

$$g'(A'_a X^*, Y^*) = g(A_a X, Y), \quad \text{for} \quad a = 1, \ldots, p.$$

Next, using (5.8), we calculate $\nabla^S_{X*}\xi^*_a$ as follows:

$$\nabla^S_{X*}\xi^*_a = -\imath' A'_a X^* + D'_{X*}\xi^*_a = -\imath' A'_a X^* + \sum_{b=1}^{p} s'_{ab}(X^*)\xi^*_b. \qquad (10.4)$$

On the other hand, using relation (9.10), we compute

$$\nabla^S_{X*}\xi^*_a = (\overline{\nabla}_X\xi_a)^* - \overline{g}(J\imath X, \xi_a)\imath'V$$
$$= (-\imath A_a X + D_X\xi_a)^* - \sum_{b=1}^{p} u^b(X)\overline{g}(\xi_b, \xi_a)\imath'V$$
$$= -\imath'(A_a X)^* + \sum_{b=1}^{p}(s_{ab}(X)\xi_b)^* - u^a(X)\imath'V, \qquad (10.5)$$

where we have put

$$J\imath X = \imath FX + \sum_{a=1}^{p} u^a(X)\xi_a. \qquad (10.6)$$

Comparing relations (10.4) and (10.5), we conclude

$$A'_a X^* = (A_a X)^* + u^a(X)V = (A_a X)^* + g(U_a, X)V, \quad (10.7)$$
$$D'_{X^*} \xi_a^* = (D_X \xi_a)^*, \quad (10.8)$$

that is, $s'_{ab}(X^*) = s_{ab}(X)^*$, where $U_a$ is defined by

$$J\xi_a = -\imath U_a + \sum_{b=1}^{p} P_{ab} \xi_b, \quad (10.9)$$

and consequently $u^a(X) = g(U_a, X)$.

Now, we consider $\nabla_V^S \xi_a^*$. Using (9.6) and (10.9) it follows

$$\nabla_V^S \xi_a^* = (J\xi_a)^* = -\imath U_a^* + \sum_{b=1}^{p} P_{ab} \xi_b^*. \quad (10.10)$$

On the other hand, from the Weingarten formula, we have

$$\nabla_V^S \xi_a^* = -\imath' A'_a V + D'_V \xi_a^*, = -\imath' A'_a V + \sum_{b=1}^{p} s'_{ab}(V) \xi_b^*. \quad (10.11)$$

Consequently, using (10.10) and (10.11), we obtain

$$A'_a V = U_a^*, \quad s'_{ab}(V) = P_{ab}, \quad (10.12)$$
$$D'_V \xi_a^* = (J\xi_a)^* + \imath U_a^*. \quad (10.13)$$

From (10.7) and (10.12), we get

$$g'(A'_a A'_b X^*, Y^*) = g(A_a A_b X, Y) + u^b(X) u^a(Y), \quad (10.14)$$

and especially,

$$g'(A'^2_a X^*, Y^*) = g(A^2_a X, Y) + u^a(X) u^a(Y). \quad (10.15)$$

For $x \in M$, let $\{e_1, \ldots, e_n\}$ be an orthonormal basis of $T_x(M)$ and $y$ be a point of $\pi^{-1}(M)$ such that $\pi(y) = x$. We take an orthonormal basis $\{e_1^*, \ldots, e_n^*, V\}$ of $T_y(\pi^{-1}(M))$. Then, using (10.12) and (10.15), we compute

$$\sum_{a=1}^{p} \text{trace } A'^2_a = \sum_{a=1}^{p} \{ \sum_{i=1}^{n} g'(A'^2_a e_i^*, e_i^*) + g'(A'^2_a V, V) \}$$

$$= \sum_{a=1}^{p} \{ \sum_{i=1}^{n} (g(A_a^2 e_i, e_i) + u^a(e_i) u^a(e_i)) + g'(A'_a V, A'_a V) \}$$

$$= \sum_{a=1}^{p} \{ \text{trace } A_a^2 + 2g(U_a, U_a) \}. \quad (10.16)$$

Summarizing, we conclude

**Proposition 10.1.** *Under the above assumptions, the following inequality*

$$\sum_{a=1}^{p} \operatorname{trace} A'^2_a \geq \sum_{a=1}^{p} \operatorname{trace} A^2_a$$

*is always valid. The equality holds, if and only if $M$ is a $J$-invariant submanifold.*

**Proposition 10.2.** *Under the conditions stated above, if $\pi^{-1}(M)$ is a totally geodesic submanifold of $\mathbf{S}^{n+p+1}$, then $M$ is a totally geodesic, $J$-invariant submanifold.*

*Proof.* Since $\pi^{-1}(M)$ is a totally geodesic submanifold of $\mathbf{S}^{n+p+1}$, using Corollary 5.1, it follows $A'_a = 0$. Relation (10.16) then implies $A_a = 0$ and $U_a = 0$, which, using relation (10.9), completes the proof. □

Further, for the normal curvature of $M$ in $\mathbf{P}^{\frac{n+p}{2}}(\mathbf{C})$, using relations (5.27) and (9.13), we obtain

$$\bar{g}(R^{\perp}(X,Y)\xi_a, \xi_b) = u^a(Y)u^b(X) - u^a(X)u^b(Y)$$
$$- 2g(FX,Y)P_{ab} + g([A_a, A_b]X, Y).$$

Therefore, if $M$ is a totally geodesic, $J$-invariant submanifold, we conclude

$$\bar{g}(R^{\perp}(X,Y)\xi_a, \xi_b) = -2g(FX,Y)P_{ab}. \tag{10.17}$$

In this case the normal space $T_x^{\perp}(M)$ is also $J$-invariant and $P_{ab}$ never vanish. We have thus proved

**Proposition 10.3.** *The normal curvature of a totally geodesic, $J$-invariant submanifold of a complex projective space never vanishes.*

This proposition shows that the normal connection of the complex projective space which is immersed standardly in a higher dimensional complex projective space, is not flat.

Finally, we give a relation between the normal curvatures $R^{\perp}$ and $R'^{\perp}$ of $M$ and $\pi^{-1}(M)$, respectively, where $M$ is an $n$-dimensional submanifold of $\mathbf{P}^{\frac{n+p}{2}}(\mathbf{C})$ and $\pi^{-1}(M)$ is the circle bundle over $M$ which is compatible with the Hopf map $\pi$. Using relation (10.14), we obtain

$$g'([A'_a, A'_b]X^*, Y^*) = g([A_a, A_b]X, Y) + u^b(X)u^a(Y) - u^b(Y)u^a(X)$$

and therefore, using relation (5.27), it follows

$$- g^S(R'^S(\imath'X^*, \imath'Y^*)\xi_a^*, \xi_b^*) + g^S(R'^{\perp}(X^*, Y^*)\xi_a^*, \xi_b^*)$$
$$= -\bar{g}(\overline{R}(\imath X, \imath Y)\xi_a, \xi_b) + \bar{g}(R^{\perp}(X,Y)\xi_a, \xi_b) + u^b(X)u^a(Y) - u^b(Y)u^a(X).$$

Using the expressions (9.12) and (9.13), for curvature tensors of $\mathbf{S}^{n+p+1}$ and $\mathbf{P}^{\frac{n+p}{2}}(\mathbf{C})$, respectively, and using relations (10.6) and (10.9), we compute

$$g^S(R'^{\perp}(X^*, Y^*)\xi_a^*, \xi_b^*) = \bar{g}(R^{\perp}(X,Y)\xi_a, \xi_b) + 2g(FX,Y)P_{ab}. \tag{10.18}$$

# 11

# Hypersurfaces of a Riemannian manifold of constant curvature

The theory of *hypersurfaces*, defined as submanifolds of codimension one, is one of the most fundamental theories of submanifolds. Therefore, in Sections 11–13 we consider hypersurfaces of a Riemannian manifold of constant curvature. This research, combined with the results obtained in Section 10, will contribute to studying real hypersurfaces of complex projective space in Section 16.

If the sectional curvature is constant for all planes $\pi$ in $T_x(\overline{M})$ and for all points $x$ of $\overline{M}$, then $\overline{M}$ is called a *space of constant curvature*. A Riemannian manifold of constant curvature is called a *space form*. Sometimes, a space form is defined as a complete simply connected Riemannian manifold of constant curvature. The following theorem due to Schur is well-known. For the proof and more details we refer the reader to [33], [68].

**Theorem 11.1.** *Let $\overline{M}$ be a connected Riemannian manifold of dimension greater than two. If the sectional curvature depends only on the point $x$, then $\overline{M}$ is a space of constant curvature.*

For a Riemannian manifold $(\overline{M}, \overline{g})$ of constant curvature $k$, the curvature tensor $\overline{R}$ of $\overline{M}$ has the following form:

$$\overline{R}(\overline{X}, \overline{Y})\overline{Z} = k\{\overline{g}(\overline{Y}, \overline{Z})\overline{X} - \overline{g}(\overline{X}, \overline{Z})\overline{Y}\}, \tag{11.1}$$

for $\overline{X}, \overline{Y}, \overline{Z} \in T(\overline{M})$.

Any space of constant curvature is locally isometric to one of the following spaces:

$$
\begin{array}{lll}
\text{Euclidean space} & \mathbf{E}^n, & (k = 0), \\
\text{sphere} & \mathbf{S}^n, & (k > 0), \\
\text{hyperbolic space} & \mathbf{H}^n, & (k < 0).
\end{array}
$$

M. Djorić, M. Okumura, *CR Submanifolds of Complex Projective Space*,
Developments in Mathematics 19, DOI 10.1007/978-1-4419-0434-8_11,
© Springer Science+Business Media, LLC 2010

Further, let $M$ be a hypersurface of $(\overline{M}, \overline{g})$ and let $\imath : M \rightarrow \overline{M}$ denote the isometric immersion. Then the Gauss formula (5.1) and Weingarten formula (5.6) reduce respectively to

$$\overline{\nabla}_X \imath Y = \imath \nabla_X Y + g(AX, Y)\xi, \tag{11.2}$$
$$\overline{\nabla}_X \xi = -\imath AX, \tag{11.3}$$

where $\xi$ is a local choice of unit normal, $X, Y \in T(M)$ and $A$ is the shape operator in the direction of $\xi$.

Consequently, denoting by $R$ the curvature tensor of a hypersurface $M$ of a Riemannian manifold $\overline{M}$ of constant curvature $k$, the Gauss equation (5.22) becomes

$$R(X, Y)Z = k\{g(Y, Z)X - g(X, Z)Y\} + g(AY, Z)AX - g(AX, Z)AY \tag{11.4}$$

and the Codazzi equation (5.23) leads to

$$(\nabla_X A)Y = (\nabla_Y A)X, \tag{11.5}$$

since

$$\begin{aligned} g((\nabla_X A)Y - (\nabla_Y A)X, Z) &= \overline{g}(\overline{R}(\imath X, \imath Y)\imath Z, \xi) \\ &= k\{\overline{g}(\imath Y, \imath Z)\overline{g}(\imath X, \xi) - \overline{g}(\imath X, \imath Z)\overline{g}(\imath Y, \xi)\} \\ &= 0. \end{aligned} \tag{11.6}$$

The eigenvalues of the shape operator $A$ are called the *principal curvatures* of the hypersurface $M$. In the following, we assume that all principal curvatures of $M$ are constant and we denote by $T_\lambda$ the eigenspace corresponding to eigenvalue $\lambda$, that is,

$$T_\lambda = \{X \in T(M) | AX = \lambda X\}.$$

**Lemma 11.1.** *For $X \in T(M)$, $Y \in T_\lambda$, $Z \in T_\mu$, it follows*

$$g((\nabla_X A)Y, Z) = (\lambda - \mu)g(\nabla_X Y, Z). \tag{11.7}$$

*Proof.* Since the shape operator $A$ is symmetric, we conclude

$$g((\nabla_X A)Y, Z) = g(\nabla_X(AY), Z) - g(A\nabla_X Y, Z) = (\lambda - \mu)g(\nabla_X Y, Z). \quad \square$$

**Lemma 11.2.** *Let $\lambda$ and $\mu$ be principal curvatures of $M$. Then we have*

(1) $\nabla_X Y \in T_\lambda$ *if $X, Y \in T_\lambda$,*

(2) $\nabla_X Y \perp T_\lambda$ *if $X \in T_\lambda$, $Y \in T_\mu$, $\lambda \neq \mu$.*

*Proof.* For any $Z \in T(M)$, by (11.5), it follows

$$
\begin{aligned}
g(A\nabla_X Y, Z) &= g((\nabla_X(AY)), Z) - g((\nabla_X A)Y, Z) \\
&= \lambda g(\nabla_X Y, Z) - g((\nabla_Z A)X, Y) \\
&= \lambda g(\nabla_X Y, Z) - g(\nabla_Z(AX), Y) + g(\nabla_Z X, AY) \\
&= \lambda g(\nabla_X Y, Z),
\end{aligned}
$$

and we conclude $A\nabla_X Y = \lambda\nabla_X Y$. This proves (1).

For $Z \in T_\lambda$, by Lemma 11.1, it follows

$$
g((\nabla_X A)Y, Z) = (\mu - \lambda)g(\nabla_X Y, Z). \tag{11.8}
$$

On the other hand, by Codazzi equation (11.6) and Lemma 11.1, we compute

$$
g((\nabla_X A)Z, Y) = g((\nabla_Z A)X, Y) = (\lambda - \mu)g(\nabla_Z X, Y). \tag{11.9}
$$

Using (1), it follows $\nabla_Z X \in T_\lambda$ for $X, Z \in T_\lambda$ and therefore $g(\nabla_Z X, Y) = 0$. Since $\nabla_X A$ is symmetric, combining relations (11.8) and (11.9), we obtain

$$
(\mu - \lambda)g(\nabla_X Y, Z) = (\lambda - \mu)g(\nabla_Z X, Y) = 0.
$$

Hence, if $\lambda \neq \mu$, then $\nabla_X Y \perp T_\lambda$. $\qquad\square$

Now we prove a theorem of E. Cartan [6], [7], [29] for a hypersurface $M^n$ whose principal curvatures are all constant.

**Theorem 11.2.** *Let $M$ be a hypersurface of a Riemannian manifold $\overline{M}$ of constant curvature $k$. Assume that $E_1, \ldots, E_n$ are local orthonormal vector fields of $M$ satisfying $AE_i = \lambda_i E_i$, with $\lambda_i$ constant. Then for every $i \in \{1, \ldots, n\}$, we have*

$$
\sum_{\substack{j=1 \\ \lambda_j \neq \lambda_i}}^{n} \frac{k + \lambda_i \lambda_j}{\lambda_i - \lambda_j} = 0. \tag{11.10}
$$

*Proof.* From the Gauss equation (11.4), it follows

$$
R(E_i, E_j)E_j = (k + \lambda_i \lambda_j)E_i. \tag{11.11}
$$

On the other hand, using the definition of a curvature tensor $R$, for $\lambda_i \neq \lambda_j$, we compute

$$
\begin{aligned}
g(R(E_i, E_j)E_j, E_i) &= g(\nabla_{E_i}\nabla_{E_j}E_j, E_i) - g(\nabla_{E_j}\nabla_{E_i}E_j, E_i) \\
&\quad - g(\nabla_{[E_i, E_j]}E_j, E_i) \\
&= g(\nabla_{E_i}E_j, \nabla_{E_j}E_i) - g(\nabla_{[E_i, E_j]}E_j, E_i),
\end{aligned}
$$

using Lemma 11.2. Hence, we conclude

$$k + \lambda_i \lambda_j = g(\nabla_{E_i} E_j, \nabla_{E_j} E_i) - g(\nabla_{[E_i,E_j]} E_j, E_i) \tag{11.12}$$

By Lemma 11.1, we get

$$g((\nabla_{[E_i,E_j]} A) E_i, E_j) = (\lambda_i - \lambda_j) g(\nabla_{[E_i,E_j]} E_i, E_j)$$
$$= (\lambda_j - \lambda_i) g(\nabla_{[E_i,E_j]} E_j, E_i)$$

from which it follows

$$g(\nabla_{[E_i,E_j]} E_j, E_i) = \frac{g((\nabla_{[E_i,E_j]} A) E_i, E_j)}{\lambda_j - \lambda_i}. \tag{11.13}$$

Now, we compute

$$g((\nabla_{[E_i,E_j]} A) E_i, E_j) = g((\nabla_{E_i} A) E_j, [E_i, E_j])$$
$$= g((\nabla_{E_i} A) E_j, \nabla_{E_i} E_j) - g((\nabla_{E_i} A) E_j, \nabla_{E_j} E_i)$$
$$= g((\nabla_{E_j} A) E_i, \nabla_{E_i} E_j) - g((\nabla_{E_i} A) E_j, \nabla_{E_j} E_i)$$
$$= (\lambda_i - \lambda_j) g(\nabla_{E_i} E_j, \nabla_{E_j} E_i),$$

that is,

$$g((\nabla_{[E_i,E_j]} A) E_j, E_i) = (\lambda_i - \lambda_j) g(\nabla_{E_i} E_j, \nabla_{E_j} E_i). \tag{11.14}$$

Combining (11.12), (11.13) and (11.14), it follows

$$k + \lambda_i \lambda_j = 2g(\nabla_{E_i} E_j, \nabla_{E_j} E_i). \tag{11.15}$$

Using Lemma 11.1, we conclude

$$g((\nabla_{E_i} A) E_j, E_s) = (\lambda_j - \lambda_s) g(\nabla_{E_i} E_j, E_s)$$

and therefore we have

$$g(\nabla_{E_i} E_j, E_s) = \frac{g((\nabla_{E_i} A) E_j, E_s)}{\lambda_j - \lambda_s}. \tag{11.16}$$

Since $\nabla_{E_i} E_j = \sum_{s=1}^n g(\nabla_{E_i} E_j, E_s) E_s$, relation (11.15) becomes

$$k + \lambda_i \lambda_j = 2 \sum_{\substack{s=1 \\ \lambda_s \neq \lambda_j}}^n g(\nabla_{E_i} E_j, E_s) g(\nabla_{E_j} E_i, E_s). \tag{11.17}$$

Substituting (11.16) into (11.17), we get

$$k + \lambda_i \lambda_j = 2 \sum_{\substack{s=1 \\ \lambda_s \neq \lambda_j, \lambda_i}}^n \frac{g((\nabla_{E_i} A) E_j, E_s)^2}{(\lambda_i - \lambda_s)(\lambda_j - \lambda_s)}.$$

Since $\nabla_{E_i} A$ is symmetric, we compute

$$\sum_{\substack{j=1 \\ \lambda_j \neq \lambda_i}}^{n} \frac{k + \lambda_i \lambda_j}{\lambda_i - \lambda_j} = -\sum_{\substack{s=1 \\ \lambda_s \neq \lambda_i}}^{n} \frac{k + \lambda_i \lambda_s}{\lambda_i - \lambda_s},$$

which proves (11.10). $\qquad\qquad\qquad\qquad\qquad\qquad\qquad\qquad\square$

As an application of Theorem 11.2, we prove the following

**Theorem 11.3.** *Let $M$ be a hypersurface of a space of nonpositive constant curvature whose principal curvatures are constant. Then at most two of them are distinct. If the ambient manifold is a Euclidean space and $M$ has two distinct principal curvatures, then one of them must be zero.*

*Proof.* First, we consider the case when the ambient manifold is a Euclidean space. Then (11.10) becomes

$$\sum_{j} \frac{\lambda_i \lambda_j}{\lambda_i - \lambda_j} = 0,$$

for any principal curvature $\lambda_i$. Now, let $\lambda_i$ be the least positive principal curvature. Then each term $\frac{\lambda_i \lambda_j}{\lambda_i - \lambda_j}$ is negative and consequently, $\lambda_j = 0$, for all $\lambda_i \neq \lambda_j$. If all principal curvatures are nonpositive, we take $\lambda_i$ in such a way that $|\lambda_i|$ is the greatest of principal curvatures and using entirely the same argument as in the above discussion, we obtain $\lambda_j = 0$ for all $\lambda_j \neq \lambda_i$.

Next, let the ambient manifold be a space of negative constant curvature. In this case we may suppose that $k = -1$. We take $\lambda_i$ in such a way that the other principal curvatures cannot be between $\lambda_i$ and $\frac{1}{\lambda_i}$. Note that we can take such $\lambda_i$, for example, to be the smallest principal curvature which is bigger than 1 or we can take the largest principal curvature satisfying $0 < \lambda_i < 1$. Then every $\frac{\lambda_i \lambda_j - 1}{\lambda_i - \lambda_j}$ is negative, unless $\lambda_j = \frac{1}{\lambda_i}$. This completes the proof. $\quad\square$

Further we consider the hypersurfaces $M$ of a Euclidean space whose principal curvatures are constant. Using Theorem 11.3, it follows that $M$ has at most two distinct principal curvatures. If the principal curvatures are all identical, $M$ is either totally geodesic or totally umbilical. In this case, if $M$ is complete, by Theorems 6.1 and 6.2, $M = \mathbf{E}^n$ or $M = \mathbf{S}^n$.

Now suppose that $M$ has two distinct principal curvatures. Then, by Theorem 11.3, one of them must be 0 and we denote by $\lambda$ the nonzero principal curvature. Since $\lambda$ is constant, the multiplicity $r$ of $\lambda$ is also constant and

$$D_\lambda = \{X \in T(M) | AX = \lambda X\},$$
$$D_0 = \{X \in T(M) | AX = 0\}$$

define distributions of dimension $r$ and dimension $n - r$, respectively. By Lemma 11.2, $D_\lambda$ and $D_0$ are both involutive and if $X \in D_\lambda$, $Y \in D_0$, then $\nabla_X Y \in D_0$, $\nabla_Y X \in D_\lambda$, which shows that $D_\lambda$ and $D_0$ are parallel along their normals in $M$.

**Lemma 11.3.** *The integral submanifolds $M_\lambda$ of $D_\lambda$ and $M_0$ of $D_0$ are both totally geodesic in $M$.*

*Proof.* For $X \in D_\lambda$ and $Y \in D_0$, we have $g(X, Y) = 0$, from which $g(\nabla_Z X, Y) + g(X, \nabla_Z Y) = 0$. If $Z \in D_\lambda$, by Lemma 11.2, $\nabla_Z Y \in D_0$ and $g(\nabla_Z X, Y) = 0$. This shows that $M_\lambda$ is totally geodesic in $M$. In entirely the same way, we can see that $M_0$ is totally geodesic. $\qquad\square$

Therefore, using the de Rham [52] decomposition theorem, we have

**Lemma 11.4.** *If $M$ is complete, $M$ is a product manifold $M_\lambda \times M_0$.*

**Lemma 11.5.** *$M_\lambda$ is totally umbilical in $\mathbf{E}^{n+1}$ and $M_0$ is totally geodesic in $\mathbf{E}^{n+1}$.*

*Proof.* Let $\imath'$ be the immersion of $M_\lambda$ into $M$. Then for any $X', Y' \in T(M_\lambda) = D_\lambda$, we have

$$\overline{\nabla}_{X'\imath} \circ \imath' Y' = \imath \circ \imath' \nabla'_{X'} Y' + \sum_{a=r+1}^{n+1} g'(A'_a X', Y') \xi'_a, \qquad (11.18)$$

where $g'$ is the induced Riemannian metric of $M_\lambda$, $\xi'_a$'s are orthonormal normals to $M_\lambda$ in $\mathbf{E}^{n+1}$ and $A'_a$ are corresponding shape operators of $\xi'_a$. Choosing a unit normal $\xi'_{n+1}$ as $\xi$, which is the unit normal to $M$ in $\mathbf{E}^{n+1}$, relation (11.18) becomes

$$\overline{\nabla}_{X'\imath} \circ \imath' Y' = \imath \circ \imath' \nabla'_{X'} Y' + \sum_{a=r+1}^{n} g'(A'_a X', Y') \xi'_a + g'(A'_{n+1} X', Y') \xi. \quad (11.19)$$

On the other hand, we have

$$\overline{\nabla}_{X'\imath} \circ \imath' Y' = \imath \nabla_{X'\imath} \imath' Y' + g(A\imath' X', \imath' Y') \xi$$

$$= \imath \{ \imath' \nabla'_{X'} Y + \sum_{a=r+1}^{n} g'(A''_a X', Y') \xi'_a \} + g(A\imath' X', \imath' Y') \xi,$$

where $A''_a$ denotes the shape operator of $M_\lambda$ with respect to $\xi'_a$ in $M$. By Lemma 11.3, $M_\lambda$ is totally geodesic in $M$, and consequently the last equation can be written as

$$\overline{\nabla}_{X'\imath} \circ \imath' Y' = \imath \circ \imath' \nabla'_{X'} Y' + \lambda g'(X', Y') \xi. \qquad (11.20)$$

Comparing (11.19) and (11.20), we have $A'_a X' = 0$ and $A'_{n+1} X' = \lambda X'$. Thus, $M_\lambda$ is a totally umbilical submanifold of $\mathbf{E}^{n+1}$. Similarly, we can prove that $M_0$ is a totally geodesic submanifold of $\mathbf{E}^{n+1}$. $\qquad\square$

By Theorems 6.1 and 6.2, $M_\lambda$ is an $r$-dimensional sphere $\mathbf{S}^r$ and $M_0$ is an $(n-r)$-dimensional Euclidean space $\mathbf{E}^{n-r}$ and $M = \mathbf{S}^r \times \mathbf{E}^{n-r}$. Thus we proved the following classification theorem.

**Theorem 11.4.** *Let $M$ be a hypersurface of an $(n+1)$-dimensional Euclidean space $\mathbf{E}^{n+1}$ whose principal curvatures are all constant. Then $M$ is one of the following:*

(1) *$n$-dimensional hypersphere $\mathbf{S}^n$;*
(2) *$n$-dimensional hyperplane $\mathbf{E}^n$;*
(3) *the product manifold of an $r$-dimensional sphere and an $(n-r)$-dimensional Euclidean space $\mathbf{S}^r \times \mathbf{E}^{n-r}$.*

# Hypersurfaces of a sphere

Here we give several examples of hypersurfaces of a sphere.

*Example* 12.1. *Small sphere.*

The hypersurface

$$M = \left\{ (y^1, \ldots, y^{n+2}) \in \mathbf{E}^{n+2} \,\middle|\, \sum_{\lambda=1}^{n+2} (y^\lambda)^2 = \frac{1}{a^2}, \quad y^{n+2} = \sqrt{\frac{1}{a^2} - \frac{1}{b^2}} \right\}$$

is called the small sphere. Then $M \subset \mathbf{S}^{n+1}(1/a) \subset \mathbf{E}^{n+2}$ and $M$ lies in a hyperplane $\mathbf{E}^{n+1}$ of $\mathbf{E}^{n+2}$ defined by

$$y^1 = x^1, \ldots, \quad y^{n+1} = x^{n+1}, \quad y^{n+2} = \sqrt{\frac{1}{a^2} - \frac{1}{b^2}}.$$

The unit normal vector field $\xi_1$ of $\mathbf{E}^{n+1}$ to $\mathbf{E}^{n+2}$ is

$$\xi_1 = \frac{\partial}{\partial y^{n+2}}.$$

From the defining equation, we have $\sum_{\lambda=1}^{n+1}(y^\lambda)^2 = \frac{1}{b^2}$ and the unit normal vector field $\xi_2$ of $M$ in $\mathbf{E}^{n+1}$ is

$$\xi_2 = b \sum_{\lambda=1}^{n+1} y^\lambda \frac{\partial}{\partial y^\lambda}.$$

Since the unit normal vector field $\xi'$ of $\mathbf{S}^{n+1}$ in $\mathbf{E}^{n+2}$ is

$$\xi' = a \sum_{\lambda=1}^{n+2} y^\lambda \frac{\partial}{\partial y^\lambda},$$

to find the unit normal vector field $\xi$ of $M$ in $\mathbf{S}^{n+1}(1/a)$, we put $\xi = \alpha\xi_2 + \beta\xi_1$. Then, from $\xi' = \frac{a}{b}\xi_2 + ay^{n+2}\xi_1$ and $\langle \xi, \xi' \rangle = 0$, we have

M. Djorić, M. Okumura, *CR Submanifolds of Complex Projective Space*, Developments in Mathematics 19, DOI 10.1007/978-1-4419-0434-8_12, © Springer Science+Business Media, LLC 2010

$$\xi = \alpha \xi_2 - \frac{\alpha}{b y^{n+2}} \xi_1.$$

Moreover $\langle \xi, \xi \rangle = 1$ implies

$$1 = \alpha^2 \left\{ b^2 \sum_{\lambda=1}^{n+1} \left( y^{\lambda} \right)^2 + \frac{1}{b^2 (y^{n+2})^2} \right\} = \frac{\alpha^2 b^2}{b^2 - a^2}.$$

Thus, $\alpha = -\frac{\sqrt{b^2 - a^2}}{b}$ and consequently

$$\xi = -\sqrt{b^2 - a^2} \left( \sum_{\lambda=1}^{n+1} y^{\lambda} \frac{\partial}{\partial y^{\lambda}} \right) + \frac{a}{b} \frac{\partial}{\partial y^{n+2}}. \tag{12.1}$$

Let $\nabla^E$ be the Euclidean connection of $\mathbf{E}^{n+2}$ and $\xi = \sum_{\lambda=1}^{n+2} \xi^{\lambda} \frac{\partial}{\partial y^{\lambda}}$. Then we have

$$\nabla^E_{\frac{\partial}{\partial x^j}} \xi = \sum_{\lambda, \mu} \frac{\partial y^{\mu}}{\partial x^j} \nabla^E_{\frac{\partial}{\partial y^{\mu}}} \left( \xi^{\lambda} \frac{\partial}{\partial y^{\lambda}} \right)$$

$$= -\sqrt{b^2 - a^2} \sum_{\mu=1}^{n+2} \frac{\partial y^{\mu}}{\partial x^j} \sum_{\lambda=1}^{n+1} \delta^{\lambda}_{\mu} \frac{\partial}{\partial y^{\lambda}} = -\sqrt{b^2 - a^2} \imath \left( \frac{\partial}{\partial x^j} \right),$$

since $\frac{\partial y^{n+2}}{\partial x^j} = 0$. Thus, using Weingarten formula (5.6), we obtain

$$\nabla^E_{\frac{\partial}{\partial x^j}} \xi = -\imath A \frac{\partial}{\partial x^j} + s \left( \frac{\partial}{\partial x^j} \right) \xi' = -\sqrt{b^2 - a^2} \imath \left( \frac{\partial}{\partial x^j} \right).$$

Hence we have, for any $X \in T(M)$,

$$AX = \sqrt{b^2 - a^2} X, \qquad s(X) = 0, \tag{12.2}$$

which means that the small sphere $M$ is a totally umbilical and not totally geodesic submanifold of the sphere $\mathbf{S}^{n+1}(1/a)$. $\diamond$

*Example* 12.2. *Product of spheres in* $\mathbf{S}^{n+1}(a)$. Let

$$M = \{ (y^1, \ldots, y^{p+1}, u^1, \ldots, u^{n-p+1}) \mid$$

$$\sum_{\lambda=1}^{p+1} (y^{\lambda})^2 = a^2 \cos^2 \theta, \quad \sum_{\lambda=1}^{n-p+1} (u^{\lambda})^2 = a^2 \sin^2 \theta \},$$

where $\theta$ is a fixed constant and $1 \leq p \leq n-1$. $M$ is the product of two spheres $M_1$ and $M_2$ where

$$M_1 = \left\{ (y^1, \ldots, y^{p+1}) \in \mathbf{E}^{p+1} \mid \sum_{\lambda=1}^{p+1} (y^{\lambda})^2 = a^2 \cos^2 \theta \right\},$$

$$M_2 = \left\{ (u^1, \ldots, u^{n-p+1}) \in \mathbf{E}^{n-p+1} \mid \sum_{\lambda=1}^{n-p+1} (u^{\lambda})^2 = a^2 \sin^2 \theta \right\}.$$

Let $\xi_1$ (respectively $\xi_2$) be the unit normal vector field to $M_1$ (respectively $M_2$) in $\mathbf{E}^{p+1}$ (respectively $\mathbf{E}^{n-p+1}$), namely,

$$\xi_1 = -\frac{1}{a\cos\theta}\sum_{\lambda=1}^{p+1}y^\lambda\frac{\partial}{\partial y^\lambda},$$

$$\xi_2 = -\frac{1}{a\sin\theta}\sum_{\lambda=1}^{n-p+1}u^\lambda\frac{\partial}{\partial u^\lambda}.$$

Now we put $y^1 = y^1, \ldots, y^{p+1} = y^{p+1}$, $y^{p+2} = u^1, \ldots, y^{n+2} = u^{n-p+1}$ and we regard that $\xi_1$ and $\xi_2$ are unit normal vector fields to $M$ in $\mathbf{E}^{n+2}$. Since the unit normal vector field $\xi'$ to $\mathbf{S}^{n+1}(a)$ in $\mathbf{E}^{n+2}$ is

$$\xi' = -\frac{1}{a}\sum_{\lambda=1}^{n+2}y^\lambda\frac{\partial}{\partial y^\lambda} = (\cos\theta)\xi_1 + (\sin\theta)\xi_2,$$

to find the unit normal vector field $\xi = \sum_{\lambda=1}^{n+2}\xi^\lambda\frac{\partial}{\partial y^\lambda}$ to $M$ in $\mathbf{S}^{n+1}(a)$, we put $\xi = \alpha\xi_1 + \beta\xi_2$ and use the conditions $\langle\xi,\xi'\rangle = 0$ and $\langle\xi,\xi\rangle = 1$. Then we have $\beta = -\cot\theta$ and $\alpha = \sin\theta$. Therefore we conclude

$$\xi = (\sin\theta)\xi_1 - (\cos\theta)\xi_2$$
$$= \frac{1}{a}\left(-\tan\theta\sum_{\lambda=1}^{p+1}y^\lambda\frac{\partial}{\partial y^\lambda} + \cot\theta\sum_{\lambda=p+2}^{n+2}y^\lambda\frac{\partial}{\partial y^\lambda}\right),$$

that is, for $1 \le \lambda \le p+1$,

$$\xi^\lambda = -\frac{1}{a}(\tan\theta)y^\lambda \tag{12.3}$$

and for $p+2 \le \lambda \le n+2$,

$$\xi^\lambda = \frac{1}{a}(\cot\theta)y^\lambda. \tag{12.4}$$

Let $(x^1, \ldots, x^n)$ denote the local coordinates of $M$. Then the immersion $\imath : M \to \mathbf{E}^{n+2}$ is represented by

$$y^1 = x^1, \quad \ldots, \quad y^p = x^p, \quad (y^{p+1})^2 = a^2\cos^2\theta - \sum_{i=1}^p(x^i)^2,$$

$$y^{p+2} = x^{p+1}, \quad \ldots, \quad y^{n+1} = x^n, \quad (y^{n+2})^2 = a^2\sin^2\theta - \sum_{i=p+1}^n(x^i)^2.$$

Using Weingarten formula (5.6), we have, for $1 \le j \le p$,

$$-\imath A\left(\frac{\partial}{\partial x^j}\right) + s\left(\frac{\partial}{\partial x^j}\right)\xi' = \sum_{\lambda,\mu=1}^{n+2}\frac{\partial y^\mu}{\partial x^j}\frac{\partial \xi^\lambda}{\partial y^\mu}\frac{\partial}{\partial y^\lambda}$$

$$= -\frac{1}{a}\tan\theta\sum_{\mu=1}^{n+2}\frac{\partial y^\mu}{\partial x^j}\frac{\partial}{\partial y^\mu}$$

$$= -\frac{1}{a}\tan\theta\,\imath\left(\frac{\partial}{\partial x^j}\right)$$

since $\frac{\partial y^\mu}{\partial x^j} = 0$ for $p+2 \le \mu \le n+2$. With the notation

$$A\frac{\partial}{\partial x^j} = \sum_{k=1}^{n}A_j^k\frac{\partial}{\partial x^k},$$

we have $A_j^k = -\frac{1}{a}\tan\theta\delta_j^k$ and $s = 0$ for $1 \le j \le p$. In entirely the same way we have $A_j^k = \frac{1}{a}\cot\theta\delta_j^k$ and $s = 0$ for $p+1 \le j \le n$. Thus the shape operator $A$ is represented by the matrix

$$A = \frac{1}{a}\begin{pmatrix} -\tan\theta I & 0 \\ 0 & \cot\theta I \end{pmatrix}$$

and all the coefficients of the third fundamental form are equal to zero. This means that the shape operator of the product of two spheres in $\mathbf{S}^{n+1}(1/a)$ has exactly two distinct principal curvatures. ◇

*Example* 12.3. [42] Let $\mathbf{S}^{2n+1}$ denote a unit sphere in $\mathbf{E}^{2n+2}$, namely,

$$\mathbf{S}^{2n+1} = \{(x_1,\ldots,x_{n+1},y_1,\ldots,y_{n+1})|\sum_{i=1}^{n+1}(x_i^2 + y_i^2) = 1\}$$

and let us consider a function

$$F(x,y) = F(x_1,\ldots,x_{n+1},y_1,\ldots,y_{n+1}) = \left\{\sum_{i=1}^{n+1}(x_i^2 - y_i^2)\right\}^2 + 4\left(\sum_{i=1}^{n+1}x_iy_i\right)^2.$$

For each $0 < \theta < \frac{\pi}{4}$, let

$$M'(2n,\theta) = \{(x,y) \in \mathbf{S}^{2n+1}|F(x,y) = \cos^2 2\theta\}$$

denote a hypersurface of $\mathbf{S}^{2n+1}$. It is well-known (see [42]) that the principal curvatures of $M'(2n,\theta)$ are $\cot(\theta - \frac{\pi}{4})$, $\cot\theta$, $\cot(\theta + \frac{\pi}{4})$, $\cot(\theta + \frac{\pi}{2}) = -\tan\theta$ with multiplicities $n-1$, $1$, $n-1$, $1$, respectively. ◇

*Remark* 12.1. R. Takagi in [57] proved that if a hypersurface $M$ of $\mathbf{S}^{2n+1}$ has four constant principal curvatures and if the multiplicity of one of them is equal to 1, then $M$ is congruent to $M'(2n,\theta)$.

It is well-known that the sphere $\mathbf{S}^{n+1}(\frac{1}{a})$ of radius $\frac{1}{a}$ is a Riemannian manifold of constant curvature $a$. Now, let $M$ be a hypersurface of $\mathbf{S}^{n+1}(\frac{1}{a}) \subset \mathbf{E}^{n+2}$ and $\xi$ be the unit normal of $M$ in $\mathbf{S}^{n+1}(\frac{1}{a})$. Let us denote with $\imath$ and $j$ the immersions $M \to \mathbf{S}^{n+1}$ and $\mathbf{S}^{n+1} \to \mathbf{E}^{n+2}$, respectively. Regarding $M$ as a submanifold of codimension 2 of $\mathbf{E}^{n+2}$, let us choose orthonormal normal vector fields $\xi_1$ and $\xi_2$ in such a way that $\xi_1 = j\,\xi$ is normal to $M$ in $\mathbf{S}^{n+1}$ and $\xi_2$ is normal to $\mathbf{S}^{n+1}$ in $\mathbf{E}^{n+2}$, that is, in the direction of the position vector field of $\mathbf{S}^{n+1}$. Then we compute

$$\nabla_X^E \xi_1 = \nabla_X^E j\xi = j\,\nabla_X^S \xi + h^S(\imath X, \xi) = -j \circ \imath\, AX - a\,g^S(\imath X, \xi) = -j \circ \imath\, AX.$$

On the other hand, using the Weingarten formula (5.6) and the notation (5.10), we obtain

$$\nabla_X^E \xi_1 = -j \circ \imath A_1 X + s(X)\xi_2.$$

Comparing the above two equations, we conclude $A_1 = A$ and $s = 0$. This means that the shape operator $A_1$ with respect to $\xi_1$ is identical with the shape operator $A$ of $M$ in $\mathbf{S}^{n+1}$. Further, if $\xi'$ is the position vector field of $\mathbf{S}^{n+1}$, then $\xi_2 = -a\,\xi'$ and therefore

$$\nabla_X^E \xi_2 = -a \nabla_X^E \xi' = -a\,j \circ \imath X.$$

On the other hand, using the Weingarten formula, we obtain

$$\nabla_X^E \xi_2 = -j \circ \imath\, A_2 X - s(X)\xi_1$$

and consequently, $A_2 = aI$ and $s = 0$. Now we prove

**Theorem 12.1.** *A totally umbilical hypersurface $M$ of $\mathbf{S}^{n+1}(\frac{1}{a})$ is a sphere or an open subset of a sphere.*

*Proof.* Under the assumptions of Theorem 12.1, using the previous notation, since $M$ is a submanifold of $\mathbf{E}^{n+2}$, the shape operators of $M$ are $A_1 = bI$ and $A_2 = aI$. Using the Codazzi equation (11.5), it follows $(Xb)Y = (Yb)X$ and therefore $b = constant$. Hence,

$$\nabla_X^E \xi_1 = -b\,j \circ \imath X,$$
$$\nabla_X^E \xi_2 = -a\,j \circ \imath X.$$

For $\xi_3 = a\xi_1 - b\xi_2$, it follows $\nabla_X^E \xi_3 = 0$, that is, $\xi_3$ is a constant vector field. Since for the position vector field $P$ of $M$ we compute

$$X\langle P, \xi_3 \rangle = \langle j \circ \imath X, \xi_3 \rangle + \langle P, \nabla_X^E \xi_3 \rangle = 0,$$

it follows $\langle P, \xi_3 \rangle = constant$. Hence $P \in M$ lies on a plane defined by $\langle P, \xi_3 \rangle = constant$ and on $\mathbf{S}^{n+1}(\frac{1}{a})$. This means that $M$ is an $n$-dimensional sphere or an open subset of it. □

# 13

## Hypersurfaces of a sphere with parallel shape operator

In [53] P. J. Ryan considered hypersurfaces of real space forms and specifically, he gave a complete classification of hypersurfaces in the sphere which satisfy a certain condition. The condition that the shape operator is parallel is its special case. In this section we give the proof of this classification (in the specific case $\nabla_X A = 0$) and furthermore, we show that the algebraic condition (13.5) on the shape operator implies that it is parallel.

Let us suppose that the shape operator $A$ of a hypersurface $M^n$ of a unit sphere is parallel. Then

$$R(X,Y)(AZ) = \nabla_X \nabla_Y (AZ) - \nabla_Y \nabla_X (AZ) - \nabla_{[X,Y]}(AZ)$$
$$= AR(X,Y)Z. \tag{13.1}$$

Using the Gauss equation (5.22) and the form of the curvature tensor of $\mathbf{S}^{n+1}(1)$, given by (11.1) for $k = 1$, we compute

$$g(Y,AZ)X - g(X,AZ)Y + g(AY,AZ)AX - g(AX,AZ)AY$$
$$= g(Y,Z)AX - g(X,Z)AY + g(AY,Z)A^2 X - g(AX,Z)A^2 Y.$$

For an orthonormal basis $\{e_i\}$, $i = 1,\ldots,n$ of $T_x(M)$ formed by the eigenvectors of $A_x$, corresponding to the eigenvalues $\lambda_i$, using (13.1), we conclude

$$(\lambda_j - \lambda_i)(\lambda_i \lambda_j + 1) = 0. \tag{13.2}$$

Under the conditions stated above, we now prove several lemmas.

**Lemma 13.1.** *For any $x \in M$, rank $A_x = 0$ or rank $A_x = n$.*

*Proof.* Assume that rank $A_x \neq n$. Then for some $i$ we have $\lambda_i = 0$ and using equation (13.2) it follows $\lambda_j = 0$. Thus all eigenvalues of $A_x$ are zero and rank $A_x = 0$. $\square$

**Lemma 13.2.** *If $A_x \neq 0$, then $A_x$ has at most two distinct eigenvalues.*

M. Djorić, M. Okumura, *CR Submanifolds of Complex Projective Space*,
Developments in Mathematics 19, DOI 10.1007/978-1-4419-0434-8_13,
© Springer Science+Business Media, LLC 2010

*Proof.* Relation (13.2) for $i = 1$ reads $(\lambda_j - \lambda_1)(\lambda_j \lambda_1 + 1) = 0$. If $\lambda_j \neq \lambda_1$, then $\lambda_j = -\frac{1}{\lambda_1}$, which shows that $A_x$ has at most two distinct eigenvalues. □

**Lemma 13.3.** *If $A$ has two distinct eigenvalues, then the multiplicities of the eigenvalues are constant.*

*Proof.* Let $\lambda$ be an eigenvalue of $A$ of multiplicity $p$ at $x \in M$ and multiplicity $q$ at $y \in M$. Then $-\frac{1}{\lambda}$ has the multiplicity $n - p$ at $x$ and $n - q$ at $y$. Therefore we compute

$$(\text{trace } A)(x) - (\text{trace } A)(y) = p\lambda(x) - q\lambda(y) - (n-p)\frac{1}{\lambda(x)} + (n-q)\frac{1}{\lambda(y)}$$

$$= (p-q)(\lambda(x) + \frac{1}{\lambda(x)}) + q(\lambda(x) - \lambda(y)) + (n-q)\frac{\lambda(x) - \lambda(y)}{\lambda(x)\lambda(y)}.$$

Since trace $A$ is continuous, this implies $p = q$.     □

Let $A$ have exactly two distinct eigenvalues $\lambda$ and $\mu(= -\frac{1}{\lambda})$. We put

$$T_\lambda(x) = \{X_x \in T_x(M)|A_x X_x = \lambda X_x\},$$
$$T_\mu(x) = \{X_x \in T_x(M)|A_x X_x = \mu X_x\}.$$

Then using Lemma 13.3, it follows that $T_\lambda(x)$ and $T_\mu(x)$ make distributions $T_\lambda$ and $T_\mu$.

**Lemma 13.4.** *The distributions $T_\lambda$ and $T_\mu$ are both involutive.*

*Proof.* Let us choose $X, Y \in T_\lambda$. Then, using Codazzi equation (11.5), it follows

$$A[X,Y] = A\nabla_X Y - A\nabla_Y X$$
$$= \nabla_X(AY) - (\nabla_X A)Y - \nabla_Y(AX) + (\nabla_Y A)X$$
$$= (X\lambda)Y - (Y\lambda)X + \lambda[X,Y].$$

Hence,

$$(A - \lambda I)[X,Y] = (X\lambda)Y - (Y\lambda)X. \tag{13.3}$$

However, the left-hand members of (13.3) belong to $T_\mu$. In fact, $[X,Y] = [X,Y]_\lambda + [X,Y]_\mu$ implies that

$$(A - \lambda I)[X,Y] = (A - \lambda I)([X,Y]_\lambda + [X,Y]_\mu)$$
$$= A[X,Y]_\lambda + A[X,Y]_\mu - \lambda[X,Y]_\lambda - \lambda[X,Y]_\mu$$
$$= (\mu - \lambda)[X,Y]_\mu \in T_\mu.$$

On the other hand, the right-hand members of (13.3) belong to $T_\lambda$ and therefore

$$A[X,Y] = \lambda[X,Y] \qquad (X\lambda)Y - (Y\lambda)X = 0. \tag{13.4}$$

This shows that the distribution $T_\lambda$ is involutive. In entirely the same way, we can see that the distribution $T_\mu$ is also involutive.     □

**Lemma 13.5.** *If the multiplicities of $\lambda$ are greater than one, then $X\lambda = 0$ and $X\mu = 0$ for $X \in T_\lambda$.*

*Proof.* If $\dim T_\lambda > 1$, we can choose $X$, $Y$ to be linearly independent. Thus $X\lambda = 0$, using (13.4). Since $\mu = -\frac{1}{\lambda}$, it follows $X\mu = -\frac{1}{\lambda^2} X\lambda = 0$, and this completes the proof. $\qquad\qquad\square$

**Theorem 13.1.** *Let $M$ be a hypersurface of $\mathbf{S}^{n+1}$ whose shape operator has exactly two distinct eigenvalues, then $M$ is locally a product of two spheres.*

*Proof.* Let $T_\lambda$ and $T_\mu$ be as above. If $X \in T_\lambda$, $Y \in T_\mu$, the Codazzi equation yields

$$\nabla_X(\mu Y) - \nabla_Y(\lambda X) = A\nabla_X Y - A\nabla_Y X.$$

Since $\lambda$ and $\mu$ are constant, we get $(A - \lambda I)\nabla_Y X = (A - \mu I)\nabla_X Y$. The left-hand side is in $T_\mu$ while the right-hand side is in $T_\lambda$. Hence both sides are zero, that is, $\nabla_Y X \in T_\lambda$, $\nabla_X Y \in T_\mu$. For $Z \in T_\lambda$,

$$g(\nabla_Z X, Y) + g(X, \nabla_Z Y) = \nabla_Z(g(X,Y)) = 0.$$

On the other hand, $\nabla_Z Y \in T_\mu$ implies $g(X, \nabla_Z Y) = 0$. Thus, we have shown $\nabla_Z X \in T_\mu^{\perp}$ for all $Z$ and $X \in T_\lambda$. Since $T_\mu^{\perp} = T_\lambda$, we may write $\nabla_{T_\lambda} T_\lambda \subset T_\lambda$ and $\nabla_{T_\lambda} T_\mu \subset T_\mu$. This means that $T_\lambda$ is a totally geodesic, parallel distribution. The same conclusion can be drawn for $T_\mu$, namely, $T_\mu$ is also a totally geodesic, parallel distribution. Hence, by de Rham decomposition theorem [52], $M$ is locally isometric to the Riemannian product of the maximal integral manifolds $M_\lambda$ and $M_\mu$.

Now we consider the integral submanifold $M_\lambda$. Let $\imath_\lambda$ be the immersion of $M_\lambda$ into $M$ and $j = \imath \circ \imath_\lambda$, that is, $j$ is the immersion of $M_\lambda$ into $\mathbf{S}^{2n+1}$ via $M$. Denoting by $h^\lambda$ and $h^\lambda_M$ the second fundamental form of $M_\lambda$ in $\mathbf{S}^{2n+1}$ and in $M$, respectively, we may calculate for any $X'$, $Y' \in T_\lambda$ the covariant derivative $\nabla^\lambda$ of $M_\lambda$ as follows:

$$\begin{aligned}
\nabla^S_{X'} jY' &= j\nabla^\lambda_{X'} Y' + h^\lambda(X', Y') \\
&= \nabla^S_{X'} \imath \circ \imath_\lambda Y' \\
&= \imath \nabla_{X'} \imath_\lambda Y' + h(\imath_\lambda X', \imath_\lambda Y') \\
&= \imath \{\imath_\lambda \nabla^\lambda_{X'} Y' + h^\lambda_M(X', Y')\} + h(\imath_\lambda X', \imath_\lambda Y') \\
&= j\nabla^\lambda_{X'} Y' + \imath h^\lambda_M h(X', Y') + h(\imath_\lambda X', \imath_\lambda Y').
\end{aligned}$$

Since $M_\lambda$ is totally geodesic in $M$, $h^\lambda_M = 0$ and we easily see that $h^\lambda(X', Y') = h(X, Y) = g(AX, Y) = \lambda g(X, Y)$. By the Gauss equation, the curvature tensor $R^\lambda$ of $M_\lambda$ satisfies

$$\begin{aligned}
g^\lambda(R^\lambda(X', Y')Z', W') &= \bar{g}(jY', jZ')\bar{g}(jY', jW') - \bar{g}(jX', jZ')\bar{g}(jY', jW') \\
&\quad + h^\lambda(Y', Z')h^\lambda(X', W') - h^\lambda(X', Z')h^\lambda(Y', W') \\
&= (1 + \lambda^2)\{g(Y, Z)g(X, W) - g(X, Z)g(Y, W)\} \\
&= (1 + \lambda^2)\{g^\lambda(Y', Z')g^\lambda(X', W') \\
&\quad - g^\lambda(X', Z')g^\lambda(Y', W')\}.
\end{aligned}$$

This shows that the integral manifold $M_\lambda$ is a Riemannian manifold of constant curvature $1 + \lambda^2$. In entirely the same way we obtain that $M_\mu$ is a Riemannian manifold of constant curvature $1 + \mu^2$. Thus, $M$ is locally a product of two spheres of radius $\frac{1}{\sqrt{1+\lambda^2}}$ and $\frac{1}{\sqrt{1+\mu^2}}$, respectively. This completes the proof.    $\square$

**Lemma 13.6.** *Let $M$ be a hypersurface of a Riemannian manifold of constant curvature $k$. If the shape operator $A$ satisfies*

$$A^2 X = \alpha\, AX + kX, \qquad X \in T(M) \tag{13.5}$$

*for some constant $\alpha$, then*

$$\nabla_X \nabla_Y A - \nabla_Y \nabla_X A - \nabla_{[X,Y]} A = R(X,Y)A = 0. \tag{13.6}$$

*Proof.* From the definition of the curvature tensor, it follows

$$
\begin{aligned}
&(\nabla_X \nabla_Y A - \nabla_Y \nabla_X A - \nabla_{[X,Y]} A)Z \\
&= (R(X,Y)A)Z = R(X,Y)(AZ) - A(R(X,Y)Z) \\
&= k\{g(Y,AZ)X - g(X,AZ)Y\} + g(AY,AZ)AX - g(AX,AZ)AY \\
&\quad - k\{g(Y,Z)AX - g(X,Z)AY\} - g(AY,Z)A^2X + g(AX,Z)A^2Y.
\end{aligned}
$$

Substituting (13.5) into the last equation, we obtain the result.    $\square$

We note that relation (13.5) implies that the eigenvalues of $A$ satisfy

$$\lambda^2 - \alpha\lambda - k = 0.$$

Since $\alpha$ and $k$ are constant, it follows that $\lambda$ is constant and therefore trace $A$, trace $A^2$ are both constant.

Let $\{e_1, \ldots, e_n\}$ be an orthonormal basis in $T_x(M)$ and extend $e_1, \ldots, e_n$ to vector fields in a neighborhood of $x$ by parallel translation along geodesics with respect to the Levi-Civita connection of $M$. Then $\nabla e_i = 0$ for $i = 1, \ldots, n$ at $x$. As $\nabla_{e_i} A$ is symmetric, using the Codazzi equation, we compute

$$
g\left(\sum_{i=1}^{n} (\nabla_{e_i} A)e_i, X\right) = \sum_{i=1}^{n} g(e_i, (\nabla_{e_i} A)X) = \sum_{i=1}^{n} g(e_i, (\nabla_X A)e_i)
$$
$$
= \text{trace } \nabla_X A = X \text{ trace } A = 0,
$$

that is,

$$\sum_{i=1}^{n} (\nabla_{e_i} A)e_i = 0. \tag{13.7}$$

Differentiating (13.7) covariantly and making use of $\nabla e_i = 0$ at $x$, we get

$$\sum_{i=1}^{n} (\nabla_X \nabla_{e_i} A)e_i = 0. \tag{13.8}$$

Let $X$ be a tangent vector at $x$ and extend it to a vector field in a normal neighborhood of $x$ by parallel translation along geodesics. Then, $\nabla X = 0$ at $x$. Therefore, at $x$, using (13.6), (13.8) and Codazzi equation, we obtain

$$\sum_{i=1}^{n}\{(\nabla_{e_i}\nabla_X A)e_i - (\nabla_X\nabla_{e_i} A)e_i - (\nabla_{[e_i,X]} A)e_i\} = \sum_{i=1}^{n}(\nabla_{e_i}\nabla_X A)e_i$$

$$= \sum_{i=1}^{n}\{\nabla_{e_i}(\nabla_X Ae_i) - (\nabla_X A)\nabla_{e_i}e_i\}$$

$$= \sum_{i=1}^{n}\nabla_{e_i}(\nabla_{e_i}AX) = \sum_{i=1}^{n}(\nabla_{e_i}\nabla_{e_i}A)X + \sum_{i=1}^{n}(\nabla_{e_i}A)(\nabla_{e_i}X)$$

$$= \sum_{i=1}^{n}(\nabla_{e_i}\nabla_{e_i}A)X = 0. \tag{13.9}$$

Since trace $A^2$ is constant, it follows

$$\frac{1}{2}YX\text{trace }A^2 = \sum_{i=1}^{n}\{g(\nabla_Y\nabla_X A)e_i, Ae_i) + g((\nabla_X A)e_i, (\nabla_Y A)e_i)\}$$

$$= \sum_{i=1}^{n}\{g((\nabla_Y\nabla_X A)e_i, Ae_i) + g((\nabla_Y A)(\nabla_X A)e_i, e_i)\} = 0.$$

Hence

$$\text{trace }(\nabla_Y A)(\nabla_X A) = -\sum_{i=1}^{n}g((\nabla_Y\nabla_X A)e_i, Ae_i)$$

and therefore trace $(\nabla_X A)^2 = -\sum_{i=1}^{n}g((\nabla_X\nabla_X A)e_i, Ae_i)$. Thus

$$g(\nabla A, \nabla A) = \sum_{i=1}^{n}\text{trace }(\nabla_{e_i}A)^2 = -\sum_{i,j=1}^{n}g((\nabla_{e_j}\nabla_{e_j}A)e_i, e_i) = 0,$$

that is, $\nabla A = 0$. This shows that we have proved the following

**Theorem 13.2.** [45] *Let $M$ be a hypersurface of a Riemannian manifold of constant curvature $k$. If the shape operator $A$ satisfies (13.5), then $\nabla A = 0$.*

# 14

## Codimension reduction of a submanifold

Let us first recall the theory of curves in 3-dimensional Euclidean space $\mathbf{E}^3$. The curve $C$, whose torsion vanishes identically, is a plane curve. In other words, for the curve $C$ without torsion, there exists a 2-dimensional totally geodesic subspace $\mathbf{E}^2$ such that $C \subset \mathbf{E}^2 \subset \mathbf{E}^3$. In general, a curve $C$ is a submanifold of codimension 2 of $\mathbf{E}^3$, but if its torsion is zero, it can be regarded as a submanifold of codimension 1 in $\mathbf{E}^2$, that is, the codimension is reduced from 2 to 1.

In this section we consider such a reduction of codimension for a general submanifold $M$ of a Euclidean space, of a sphere and of a complex projective space. Moreover, we give some sufficient conditions for the existence of totally geodesic submanifolds $M'$ of the ambient manifolds $\overline{M}$ such that

$$M \subset M' \subset \overline{M}.$$

For an $n$-dimensional submanifold $M$ of an $(n+p)$-dimensional Riemannian manifold $(\overline{M}, \overline{g})$, the *first normal space* $N_1(x)$ is defined to be the orthogonal complement of $\{\xi \in T_x^\perp(M) | A_\xi = 0\}$ in $T_x^\perp(M)$. Let $\overline{M}$ be a Riemannian manifold of constant curvature $k$. Then the curvature tensor $\overline{R}$ of $\overline{M}$ is given by (11.1) and using (5.27), the normal curvature $R^\perp$ of $M$ is given by

$$\overline{g}(R^\perp(X,Y)\xi_a, \xi_b) = g([A_a, A_b]X, Y). \tag{14.1}$$

**Lemma 14.1.** *Suppose that the first normal space $N_1(x)$ is invariant under parallel translation with respect to the normal connection and that the dimension of $N_1(x)$ is constant. Let $N_0(x)$ be the orthogonal complement of $N_1(x)$ in $T_x^\perp(M)$ and for $x \in M$, let $S(x) = T_x(M) + N_1(x)$. Then for any $x \in M$ there exist differentiable orthonormal vector fields $\xi_1, \ldots, \xi_p$ defined in a neighborhood $U(x)$ of $x$ such that*

*(1) for any $y \in U(x)$, $\xi_1(y), \ldots, \xi_q(y)$ span $N_1(y)$ and $\xi_{q+1}(y), \ldots, \xi_p(y)$ span $N_0(y)$;*

*(2) $\overline{\nabla}_X \xi_a = 0$ in $U(x)$ for $a \geq q+1$;*

M. Djorić, M. Okumura, *CR Submanifolds of Complex Projective Space*, Developments in Mathematics 19, DOI 10.1007/978-1-4419-0434-8_14, © Springer Science+Business Media, LLC 2010

(3) $S(y)$, for $y \in U(x)$, is invariant under parallel translation with respect to the connection in $\overline{M}$ along any curve in $U(x)$.

*Proof.* Since $N_1(x)$ is invariant under parallel translation with respect to the normal connection $D$, it follows that if $\xi \in N_1$, then $D_X \xi \in N_1$. Hence, we have for $\xi \in N_1$, $\eta \in N_0$,

$$\overline{g}(D_X \eta, \xi) = X\overline{g}(\eta, \xi) - \overline{g}(\eta, D_X \xi) = 0,$$

which shows that $N_0$ is also invariant under the parallel translation with respect to the normal connection. At $x \in M$ we choose orthonormal vectors $\xi'_1(x), \ldots, \xi'_p(x)$ in such a way that $\xi'_1(x), \ldots, \xi'_q(x)$ span $N_1(x)$ and $\xi'_{q+1}(x), \ldots, \xi'_p(x)$ span $N_0(x)$. Extending $\xi'_1, \ldots, \xi'_p$ to differentiable orthonormal normal vector fields defined in a neighborhood $U(x)$ by parallel translation with respect to the normal connection along geodesics in $M$, proves (1).

Let $\xi'_1, \ldots, \xi'_p$ be chosen as in (1). Since both $N_1$ and $N_0$ are invariant with respect to the normal connection, it follows $s'_{ab} = 0$, for $a \geq q+1$, $b \leq q$. The skew symmetric property of $s'_{ab}$ implies $s'_{ab} = 0$, for $a \leq q$, $b \geq q+1$. Moreover, if $a \geq q+1$, then $\xi'_a \in N_0$ and consequently $A'_a = 0$. Hence, for $b = 1, \ldots, p$, it follows $[A'_a, A'_b]X = 0$. This, together with (14.1), implies $R^\perp(X, Y)\xi'_a = 0$, for $a \geq q+1$. In entirely the same way, we can choose local orthonormal normal vector fields $\xi_1, \ldots, \xi_p$ in such a way that $\xi_{q+1}, \ldots, \xi_p \in \text{span}\{\xi'_{q+1}, \ldots, \xi'_p\}$ and $s_{ab} = 0$ for $a, b \geq q+1$. This proves (2).

To prove (3) it suffices to show that $\overline{\nabla}_X Z \in S$ whenever $Z \in S$ and $X$ is tangent to $M$. This follows from (5.1), (5.6) and (1), (2) above.  □

Now, let $\overline{M}$ be an $(n+p)$-dimensional Euclidean space.

**Theorem 14.1.** [28] *Let $M$ be an $n$-dimensional submanifold of an $(n+p)$-dimensional Euclidean space $\mathbf{E}^{n+p}$. If the first normal space $N_1$ is invariant under parallel translation with respect to the connection in the normal bundle and $q$ is the constant dimension of $N_1$, then there exists a totally geodesic $(n+q)$-dimensional submanifold $\mathbf{E}^{n+q}$ of $\mathbf{E}^{n+p}$ such that $M \subset \mathbf{E}^{n+q}$.*

*Proof.* For $x \in M$ let us define $\xi_1, \ldots, \xi_p$ and $U(x)$ as in Lemma 14.1. Further, let us define functions $f_a = \langle \mathbf{x}, \xi_a \rangle$, where $\mathbf{x}$ is the position vector and $\langle , \rangle$ is the Euclidean inner product. Then

$$X f_a = \overline{\nabla}_X f_a = \langle \imath X, \xi_a \rangle + \langle \mathbf{x}, \overline{\nabla}_X \xi_a \rangle = 0$$

for $a \geq q+1$ and $X$ tangent to $U(x)$. Thus, for $\mathbf{x} = \sum_{i=1}^{n+p} x^i e_i$ and $\xi_a = \sum_{i=1}^{n+p} \xi_a^i e_i$, we have

$$\langle \mathbf{x}, \xi_a \rangle = x^1 \xi_a^1 + \cdots + x^{n+p} \xi_a^{n+p} = c_a,$$

for $a \geq q+1$ where $c_a$ are constants. This shows that $U(x)$ lies in the intersection of $p-q$ hyperplanes, whose normal vectors are linearly independent, and the desired result is true locally. That is, for $x \in M$, there exist a neighborhood $U(x)$ of $x$ and a Euclidean subspace $\mathbf{E}^{n+q}$ such that $U(x) \subset \mathbf{E}^{n+q}$.

To get the global result we use the connectedness of $M$. Let $x$, $y \in M$ with neighborhoods $U(x)$ and $U(y)$, respectively, such that $U(x) \cap U(y) \neq \emptyset$ and $U(x) \subset \mathbf{E}_1^{n+q}$, $U(y) \subset \mathbf{E}_2^{n+q}$. Then $U(x) \cap U(y) \subset \mathbf{E}_1^{n+q} \cap \mathbf{E}_2^{n+q}$. If $\mathbf{E}_1^{n+q} \neq \mathbf{E}_2^{n+q}$, then $\mathbf{E}_1^{n+q} \cap \mathbf{E}_2^{n+q} = \mathbf{E}^{n+r}$, $r < q$, and this implies that $\dim N_1(x) < q$ for $z \in U(x) \cap U(y)$. Since $\dim N_1 = constant = q$, we conclude $\mathbf{E}_1^{n+q} = \mathbf{E}_2^{n+q}$, which proves the global result.                    $\square$

**Theorem 14.2.** *[28] Let $M$ be an $n$-dimensional submanifold of an $(n+p)$-dimensional sphere $\mathbf{S}^{n+p}$. If the first normal space $N_1(x)$ is invariant under parallel translation with respect to the connection in the normal bundle and $q$ is the constant dimension of $N_1$, then there exists a totally geodesic submanifold $\mathbf{S}^{n+q}$ of $\mathbf{S}^{n+p}$ of dimension $n+q$ such that $M \subset \mathbf{S}^{n+q}$.*

*Proof.* Let $\imath$ be the immersion of $M$ into $\mathbf{S}^{n+p}$ and let us consider $\mathbf{S}^{n+p}$ as the unit sphere in $\mathbf{E}^{n+p+1}$ with center at the origin of $\mathbf{E}^{n+p+1}$. Denoting by $\imath'$ the immersion $\mathbf{S}^{n+p} \to \mathbf{E}^{n+p+1}$, we can regard $M$ as a submanifold of $\mathbf{E}^{n+p+1}$ with the immersion $j = \imath' \circ \imath$. Let $\eta$ be the inward unit normal of $\mathbf{S}^{n+p}$, $\overline{N}_1(x)$ the first normal space for $M$ in $\mathbf{E}^{n+p+1}$, $\overline{\nabla}$ the Euclidean connection in $\mathbf{E}^{n+p+1}$, and let $\xi_1, \ldots, \xi_p$ be chosen as in Lemma 14.1. Then, $\eta_a = \imath' \xi_a$, $a = 1, \ldots, p$ and $\eta$ are mutually orthonormal normals to $M$ in $\mathbf{E}^{n+p+1}$. We note that $\overline{\nabla}_X \eta = -jX$ and, using the Gauss formula (5.1), we compute

$$\overline{\nabla}_X \eta_a = \overline{\nabla}_X \imath' \xi_a = \imath' \nabla_X \xi_a - \langle \imath X, \xi_a \rangle \eta = \imath' \nabla_X \xi_a$$

for $X$ tangent to $M$. From this it follows

$$\overline{N}_1(x) = \imath' N_1(x) + \text{span} \{\eta(x)\}$$

and therefore, $\overline{N}_1$ is invariant under the parallel translation with respect to the normal connection, where $M$ is viewed as immersed in $\mathbf{E}^{n+p+1}$. Thus, by Theorem 14.1, there exists a totally geodesic subspace $\mathbf{E}^{n+q+1}$ such that $M \subset \mathbf{E}^{n+q+1}$, that is,

$$\mathbf{E}^{n+q+1} = jT_x(M) + \imath' N_1(x) + \text{span} \{\eta(x)\},$$

for any $x \in M$. Hence $\mathbf{E}^{n+q+1}$ contains $\eta$ and therefore passes through the origin of $\mathbf{E}^{n+p+1}$. Thus

$$M \subset \mathbf{E}^{n+q+1} \cap \mathbf{S}^{n+p}(1) = \mathbf{S}^{n+q}(1),$$

which completes the proof.                    $\square$

Let $M$ be an $n$-dimensional submanifold of a complex projective space $\mathbf{P}^{\frac{n+p}{2}}(\mathbf{C})$ with complex structure $J$. For any $X \in T(M)$ and $\xi \in T^{\perp}(M)$, $J\imath X$ and $J\xi$ are written as sums of the tangential and the normal parts in the following way:

$$J\imath X = \imath F X + v(X), \quad J\xi = -\imath U_\xi + P\xi.$$

For the subspace

$$N_0(x) = \{\xi \in T_x^\perp(M)|A_\xi = 0\}$$

of $T_x^\perp(M)$, we put

$$H_0(x) = JN_0(x) \cap N_0(x).$$

Then $H_0(x)$ is the maximal $J$-invariant subspace of $N_0(x)$ and $JH_0(x) = H_0(x)$, since $J$ is an isomorphism. Moreover, we can easily conclude

**Proposition 14.1.** *For any $\xi \in H_0(x)$, it follows $A_\xi = 0$ and $U_\xi = 0$.*

Further, we denote by $H_1(x)$ the orthogonal complement of $H_0(x)$ in $T_x^\perp(M)$. By definition, the first normal space $N_1(x)$ is a subspace of $H_1(x)$ and we have

**Proposition 14.2.** *If $M$ is a complex submanifold of a Kähler manifold, then $H_1(x) = N_1(x)$.*

*Proof.* Since $H_1(x)$ and $N_1(x)$ are the orthogonal complements of $H_0(x)$ and $N_0(x)$ respectively, it only remains to verify that $H_0(x) = N_0(x)$.

Using the Weingarten formula (5.6), it follows

$$\overline{\nabla}_X(J\xi) = J\overline{\nabla}_X\xi = J(-\imath A_\xi X + D_X\xi) = -J\imath A_\xi X + JD_X\xi, \qquad (14.2)$$

since $J$ is covariantly constant. On the other hand, $M$ being a complex submanifold, $T_x(M)$ is $J$-invariant and so is $T_x^\perp(M)$, that is, for any $\xi \in T_x^\perp(M)$, it follows $J\xi \in T_x^\perp(M)$. Hence we have

$$\overline{\nabla}_X(J\xi) = -\imath A_{J\xi} X + D_X(J\xi). \qquad (14.3)$$

Comparing the tangential part and the normal part of equations (14.2) and (14.3), we conclude

$$A_{J\xi} X = J\imath A_\xi X.$$

Thus, if $\xi \in N_0(x)$, then $A_{J\xi} = 0$ and $\xi \in JN_0(x)$. This shows that $\xi \in H_0(x)$, which completes the proof. $\qquad \square$

**Proposition 14.3.** *Let $H(x)$ be a $J$-invariant subspace of $H_0(x)$ and $H_2(x)$ be the orthogonal complement of $H(x)$ in $T_x^\perp(M)$. Then $T_x(M) \oplus H_2(x)$ is a $J$-invariant subspace of $T_{\imath(x)}(\overline{M})$.*

*Proof.* Note that

$$T_{\imath(x)}(\overline{M}) = \imath T_x(M) \oplus H_2(x) \oplus H(x).$$

Under the assumptions of a proposition, $H(x)$ is $J$-invariant, that is $JH(x) = H(x)$. Thus, for any $\xi \in H(x)$ there exists $\eta \in H(x)$ such that $J\eta = \xi$. Let $Y \in \imath T_x(M) \oplus H_2(x)$. Then for $\xi \in H(x)$, it follows

$$\bar{g}(JY, \xi) = \bar{g}(JY, J\eta) = \bar{g}(Y, \eta) = 0.$$

Consequently, $JY \in \imath T_x(M) \oplus H_2(x)$ and hence, $\imath T_x(M) \oplus H_2(x)$ is a $J$-invariant subspace of $T_x(\overline{M})$.    □

For $x' \in \pi^{-1}(M)$, let

$$N_0'(x') = \{\xi' \in T_{x'}^{\perp}(\pi^{-1}(M)) | A_{\xi'}' = 0\},$$

where $A_{\xi'}'$ is the shape operator with respect to $\xi'$. Then we have

**Lemma 14.2.** *If $x'$ is a point such that $\pi(x') = x$, then*

$$N_0'(x') = \mathrm{span}\{\xi_a^* | A_a = 0, U_a = 0\},$$

*where $\xi_a \in T_x^{\perp}(M)$ and $\xi_a^*$ is the horizontal lift of $\xi_a$ at $x'$.*

*Proof.* We note that using the horizontal lift $X^*$ of some $X \in T_x(M)$, any tangent vector $X' \in T_{x'}(\pi^{-1}(M))$ can be decomposed as $X' = X^* + \alpha V$. From (10.7) and the first equation of (10.12), it follows

$$A_a' X' = A_a' X^* + \alpha A_a' V = (A_a X)^* - g(U_a, X)V - \alpha U_a^*.$$

If $A_a = 0$ and $U_a = 0$ hold, then $A_a' X' = 0$, which implies $\xi_a^* \in N_0'(x')$. Conversely, if $\xi_a^* \in N_0'(x')$, we conclude

$$(A_a X)^* - \alpha U_a^* = g(U_a, X)V.$$

In the last equation, the left-hand side member is vertical. Hence, $g(U_a, X) = 0$ for any $X \in T_x(M)$ and therefore $U_a = 0$, $A_a = 0$, which completes the proof.    □

Now we prove

**Theorem 14.3.** [48] *Let $M$ be an $n$-dimensional real submanifold of a real $(n + p)$-dimensional complex projective space $\mathbf{P}^{\frac{n+p}{2}}(\mathbf{C})$ and $H(x)$ be a $J$-invariant subspace of $H_0(x)$. If the orthogonal complement $H_2(x)$ of $H(x)$ in $T_x^{\perp}(M)$ is invariant under parallel translation with respect to the normal connection and $q$ is the constant dimension of $H_2$, then there exists a real $(n+q)$-dimensional totally geodesic complex projective subspace $\mathbf{P}^{\frac{n+q}{2}}(\mathbf{C})$ such that $M \subset \mathbf{P}^{\frac{n+q}{2}}(\mathbf{C})$.*

*Proof.* We first prove that if $M$ satisfies the conditions of Theorem 14.3, then $\pi^{-1}(M)$ satisfies the conditions of Theorem 14.2.

For $\xi_a \in H(x)$, it follows $\xi_a \in H_0(x)$ and, by Proposition 14.1, we conclude that $A_a = 0$ and $U_a = 0$ which, together with (10.7) and (10.9), implies $A_a' = 0$. This shows that, for a point $x'$ such that $\pi(x') = x$,

$$H(x)^* = \{\xi^* | \xi \in H(x)\}$$

is a subspace of $N_0'(x')$. Hence, the orthogonal complement

$$H_2(x)^* = \{\xi^* | \xi \in H_2(x)\}$$

of $H(x)^*$ in $T_{x'}^{\perp}(\pi^{-1}(M))$ is a subspace of $T_{x'}^{\perp}(\pi^{-1}(M))$ such that $N_1'(x') \subset H_2(x)^*$.

Since $H_2(x)$ is invariant under parallel translation with respect to the normal connection $D$, so is $H(x)$. This means that $D_X \xi \in H(x)$ holds for any $\xi \in H(x)$. Thus, from (10.8), (10.13) and Proposition 14.1, we conclude

$$D_{X_*}'\xi^* = (D_X \xi)^* \in H(x)^*,$$
$$D_V'\xi^* = -(J\xi)^* \in H(x)^*,$$

where $D'$ is the normal connection of $\pi^{-1}(M)$ in $\mathbf{S}^{n+p+1}$. Hence $H(x)^*$ is invariant under parallel translation with respect to the normal connection of $\pi^{-1}(M)$ in $\mathbf{S}^{n+p+1}$.

Theorem 14.2 now implies that there exists a totally geodesic submanifold $\mathbf{S}^{n+q+1}$ of $\mathbf{S}^{n+p+1}$ such that $\pi^{-1}(M) \subset \mathbf{S}^{n+q+1}$. Let $U(x')$ be a neighborhood of such a point $x'$ that $\pi^{-1}(x') = x$. Then the tangent space $T_{y'}(\mathbf{S}^{n+q+1})$ of a totally geodesic submanifold at $y' \in U(x')$ is

$$T_{y'}(\pi^{-1}(M)) \oplus H_2(y)^* = (T_y(M) \oplus H_2(y))^* \oplus \imath V,$$

where $y = \pi(y')$.

For the geodesic $\gamma$ in the direction of $V$, $\imath\gamma$ is also a geodesic of $\mathbf{S}^{n+p+1}$, since $\mathbf{S}^{n+q+1}$ is a totally geodesic submanifold. Thus, $\gamma$ is a great circle on a unit sphere $\mathbf{S}^{n+q+1}$. Since $\imath V = V'$, the Hopf fibration $\mathbf{S}^{n+q+1} \to \mathbf{P}^{\frac{n+q}{2}}(\mathbf{C})$ by $\gamma$ is compatible with the Hopf fibration $\pi : \mathbf{S}^{n+p+1} \to \mathbf{P}^{\frac{n+p}{2}}(\mathbf{C})$ and the tangent space of $\mathbf{P}^{\frac{n+q}{2}}(\mathbf{C})$ at $x$ is $T_x(M) \oplus H_2(x)$. Moreover, by Proposition 14.3, $\mathbf{P}^{\frac{n+q}{2}}(\mathbf{C})$ is a $J$-invariant subspace of $\mathbf{P}^{\frac{n+p}{2}}(\mathbf{C})$, which completes the proof.                                                     $\square$

# 15

# CR submanifolds of maximal CR dimension

In this section we continue our study of CR submanifolds of complex manifolds in the special case when the CR dimension is maximal. Having in mind Proposition 7.8, let us suppose that the ambient space is a complex manifold $(\overline{M}^{\frac{n+p}{2}}, J)$ equipped with a Hermitian metric $\overline{g}$. If $M$ is an $n$-dimensional CR submanifold of maximal CR dimension of $\overline{M}^{\frac{n+p}{2}}$, then at each point $x$ of $M$, the real dimension of $JT_x(M) \cap T_x(M)$ is $n - 1$. Therefore $M$ is necessarily odd-dimensional and there exists a unit vector $\xi_x$ normal to $T_x(M)$ such that

$$JT_x(M) \subset T_x(M) \oplus \text{span}\{\xi_x\}, \quad \text{for any} \quad x \in M. \tag{15.1}$$

Hence, for any $X \in T(M)$, we may write

$$J\imath X = \imath FX + u(X)\xi, \tag{15.2}$$

where $F$ is an endomorphism acting on $T(M)$ and $u$ is one-form on $M$.

Since $\overline{g}$ is a Hermitian metric, $J$ is skew-symmetric and therefore, using (15.2) we compute

$$g(FX, Y) = \overline{g}(J\imath X, \imath Y) = -\overline{g}(\imath X, J\imath Y)$$
$$= -\overline{g}(\imath X, \imath FY + u(Y)\xi) = -g(X, FY).$$

Hence, $F$ is a skew-symmetric endomorphism acting on $T(M)$.

Now, assume that $\eta$ is an element of $T^{\perp}(M)$ which is orthogonal to $\xi$. Then, for any $X \in T(M)$, using the Hermitian property (4.1) and (15.2), we conclude

$$\overline{g}(J\eta, \imath X) = -\overline{g}(\eta, J\imath X) = -\overline{g}(\eta, \imath FX + u(X)\xi) = 0, \tag{15.3}$$

which shows that $J\eta \in T^{\perp}(M)$. On the other hand, using (4.1), (15.2) and (15.3), we obtain

$$0 = \overline{g}(\imath X, \eta) = \overline{g}(J\imath X, J\eta) = \overline{g}(\imath FX, J\eta) + u(X)\overline{g}(\xi, J\eta) = u(X)\overline{g}(\xi, J\eta).$$

M. Djorić, M. Okumura, *CR Submanifolds of Complex Projective Space*,
Developments in Mathematics 19, DOI 10.1007/978-1-4419-0434-8_15,
© Springer Science+Business Media, LLC 2010

Let us suppose that $u_x(X) = 0$, for any $X \in T(M)$, at a point $x \in M$. Then, using (15.2), we conclude

$$J\imath X = \imath FX \quad \text{for all} \quad X \in T(M).$$

Thus $T_x(M)$ is $J$-invariant and consequently $M$ is even-dimensional (see Example 7.2), which is a contradiction. Therefore we deduce

$$\bar{g}(\xi, J\eta) = 0. \tag{15.4}$$

This means that $J\eta \perp T(M) \oplus \text{span}\,\{\xi\}$. In other words, the subbundle

$$T_1^\perp(M) = \{\eta \in T^\perp(M)|\, \bar{g}(\eta, \xi) = 0\}$$

is $J$-invariant and this result will prove extremely useful. Summarizing we have

**Lemma 15.1.** *The subbundle $T_1^\perp(M)$ is $J$-invariant and we can choose a local orthonormal basis of $T^\perp(M)$ in the following way:*

$$\xi, \xi_1, \ldots, \xi_q, \xi_{1^*}, \ldots, \xi_{q^*},$$

*where $\xi_{a^*} = J\xi_a$, $a = 1, \ldots, q$ and $q = \frac{p-1}{2}$.*

Moreover, since using (4.1) and (15.4) it follows $\bar{g}(J\xi, \eta) = -\bar{g}(\xi, J\eta) = 0$, we conclude

$$J\xi = -\imath U. \tag{15.5}$$

Now, applying $J$ to (15.2), (15.5) and comparing the tangential part and the normal part to $M$, we deduce

$$F^2 X = -X + u(X)U, \tag{15.6}$$
$$u(FX) = 0, \quad FU = 0. \tag{15.7}$$

A differentiable manifold $M'$ is said to have an *almost contact structure* if it admits a vector field $U$, a one-form $u$ and a $(1,1)$-tensor field $F$ satisfying (15.6) and (15.7). The tensor field $F$ is called the *almost contact tensor field*. In this sense, a CR submanifold of maximal CR dimension is equipped with an almost contact structure which is naturally induced from the almost complex structure of the complex manifold $(\overline{M}^{\frac{n+p}{2}}, J)$.

*Remark* 15.1. The fact that the real hypersurface (resp. submanifold) of a complex manifold which satisfies relation (15.1) admits a naturally induced almost contact structure was first announced by Tashiro [58] (resp. [59]).

Denoting by $g$ the induced Riemannian metric from the Hermitian metric $\bar{g}$ to $M$, using (15.2) and (15.5), we compute

$$g(U,U) = \overline{g}(J\xi, J\xi) = \overline{g}(\xi,\xi) = 1,$$
$$g(X,U) = \overline{g}(\imath X, \imath U) = -\overline{g}(\imath X, J\xi) = \overline{g}(J\imath X, \xi)$$
$$= \overline{g}(\imath FX, \xi) + u(X)\overline{g}(\xi,\xi) = u(X),$$

namely,

$$g(U,U) = 1, \tag{15.8}$$

$$g(U,X) = u(X). \tag{15.9}$$

Further, let us denote by $\overline{\nabla}$ and $\nabla$ the Riemannian connections of $\overline{M}$ and $M$, respectively, and by $D$ the normal connection induced from $\overline{\nabla}$ in the normal bundle $T^{\perp}(M)$. Using Lemma 15.1, we can write

$$D_X\xi = \sum_{a=1}^{q}\{s_a(X)\xi_a + s_{a^*}(X)\xi_{a^*}\} \tag{15.10}$$

and the following lemma holds:

**Lemma 15.2.** *Under the above notation, for a CR submanifold of maximal CR dimension, the vector field $\xi$ is parallel with respect to the normal connection $D$, if and only if $s_a = s_{a^*} = 0$, for a.$= 1,\ldots,q$.*

Moreover, using the basis constructed in Lemma 15.1, the Weingarten formulae can be written as follows:

$$\overline{\nabla}_X\xi = -\imath AX + D_X\xi \tag{15.11}$$

$$= -\imath AX + \sum_{a=1}^{q}\{s_a(X)\xi_a + s_{a^*}(X)\xi_{a^*}\},$$

$$\overline{\nabla}_X\xi_a = -\imath A_a X + D_X\xi_a = -\imath A_a X - s_a(X)\xi \tag{15.12}$$

$$+ \sum_{b=1}^{q}\{s_{ab}(X)\xi_b + s_{ab^*}(X)\xi_{b^*}\},$$

$$\overline{\nabla}_X\xi_{a^*} = -\imath A_{a^*}X + D_X\xi_{a^*} \tag{15.13}$$

$$= -\imath A_{a^*}X - s_{a^*}(X)\xi + \sum_{b=1}^{q}\{s_{a^*b}(X)\xi_b + s_{a^*b^*}(X)\xi_{b^*}\},$$

where $A$, $A_a$, $A_{a^*}$ are the shape operators for the normals $\xi$, $\xi_a$, $\xi_{a^*}$, respectively, and $s$'s are the coefficients of the normal connection $D$. They are related to the second fundamental form by

$$h(X,Y) = g(AX,Y)\xi \tag{15.14}$$

$$+ \sum_{a=1}^{q}\{g(A_a X,Y)\xi_a + g(A_{a^*}X,Y)\xi_{a^*}\}.$$

When the ambient complex manifold $(\overline{M}, J)$ is a Kähler manifold, using Theorem 4.2, it follows $\overline{\nabla} J = 0$. Therefore, taking the covariant derivative of $\xi_{a^*} = J\xi_a$, and using (15.2), (15.5), (15.12) and (15.13), we compute

$$A_{a^*}X = FA_aX - s_a(X)U, \tag{15.15}$$

$$A_aX = -FA_{a^*}X + s_{a^*}(X)U, \tag{15.16}$$

$$s_{a^*}(X) = u(A_aX) = g(A_aU, X), \tag{15.17}$$

$$s_a(X) = -u(A_{a^*}X) = -g(A_{a^*}U, X), \tag{15.18}$$

$$s_{a^*b^*} = s_{ab}, \quad s_{a^*b} = -s_{ab^*}, \tag{15.19}$$

for all $X, Y$ tangent to $M$ and $a, b = 1, \ldots, q$.

Further, since $F$ is skew-symmetric and $A_a$, $A_{a^*}$, $a = 1, \ldots, q$ are symmetric, using relations (15.15) and (15.16), we compute

$$\text{trace } A_{a^*} = \sum_{i=1}^{n} g(A_{a^*}e_i, e_i)$$

$$= \sum_{i=1}^{n} \{g(FA_ae_i, e_i) - s_a(e_i)g(U, e_i)\}$$

$$= \text{trace } FA_a - s_a(U) = -s_a(U),$$

$$\text{trace } A_a = s_{a^*}(U),$$

namely,

$$\text{trace } A_a = s_{a^*}(U), \quad \text{trace } A_{a^*} = -s_a(U), \quad \text{for} \quad a = 1, \ldots, q. \tag{15.20}$$

Moreover, relations (15.15)–(15.18) imply

$$g((A_aF + FA_a)X, Y) = u(Y)s_a(X) - u(X)s_a(Y), \tag{15.21}$$

$$g((A_{a^*}F + FA_{a^*})X, Y) = u(Y)s_{a^*}(X) - u(X)s_{a^*}(Y), \tag{15.22}$$

for all $a = 1, \ldots, q$.

If the vector field $\xi$ is parallel with respect to the normal connection $D$, using Lemma 15.2 and relations (15.15)–(15.18), we conclude

$$A_aU = 0, \qquad A_{a^*}U = 0, \tag{15.23}$$

$$A_aX = -FA_{a^*}X, \quad A_{a^*}X = FA_aX, \tag{15.24}$$

for all $X$ tangent to $M$ and all $a = 1, \ldots, q$.

Further, we differentiate (15.2) and (15.5) covariantly and compare the tangential part and the normal part. Then we obtain

$$(\nabla_X F)Y = u(Y)AX - g(AY, X)U, \tag{15.25}$$

$$(\nabla_Y u)(X) = g(FAY, X), \tag{15.26}$$

$$\nabla_X U = FAX. \tag{15.27}$$

Now, let us suppose, for the moment, that the ambient manifold $\overline{M}$ is a complex space form that is a Kähler manifold of constant holomorphic sectional curvature $4k$. Then the curvature tensor $\overline{R}$ of $\overline{M}$ satisfies (9.21). Consequently, using (15.2), the Gauss equation (5.22) and the Codazzi equation (5.23) for the normal $\xi$ become

$$
\begin{aligned}
R(X,Y)Z = {} & k\,\{g(Y,Z)X - g(X,Z)Y + g(FY,Z)FX \\
& - g(FX,Z)FY - 2g(FX,Y)FZ\} \\
& + g(AY,Z)AX - g(AX,Z)AY \\
& + \sum_{b=1}^{q}\Big\{g(A_bY,Z)A_bX - g(A_bX,Z)A_bY\Big\} \\
& + \sum_{b=1}^{q}\Big\{g(A_{b^*}Y,Z)A_{b^*}X - g(A_{b^*}X,Z)A_{b^*}Y\Big\},
\end{aligned}
\tag{15.28}
$$

$$
\begin{aligned}
(\nabla_X A)Y - (\nabla_Y A)X = {} & k\,\Big\{u(X)FY - u(Y)FX - 2g(FX,Y)U\Big\} \\
& + \sum_{b=1}^{q}\Big\{s_b(X)A_bY - s_b(Y)A_bX\Big\} \\
& + \sum_{b=1}^{q}\Big\{s_{b^*}(X)A_{b^*}Y - s_{b^*}(Y)A_{b^*}X\Big\},
\end{aligned}
\tag{15.29}
$$

for all $X, Y, Z$ tangent to $M$, where $R$ denotes the Riemannian curvature tensor of $M$. Moreover, the Codazzi equations for normal vectors $\xi_a$, $\xi_{a^*}$, respectively, become

$$
\begin{aligned}
(\nabla_X A_a)Y - (\nabla_Y A_a)X = {} & s_a(Y)AX - s_a(X)AY \\
& + \sum_{b=1}^{q}\Big\{s_{ab}(X)A_bY - s_{ab}(Y)A_bX\Big\} \\
& + \sum_{b=1}^{q}\Big\{s_{ab^*}(X)A_{b^*}Y - s_{ab^*}(Y)A_{b^*}X\Big\},
\end{aligned}
\tag{15.30}
$$

$$
\begin{aligned}
(\nabla_X A_{a^*})Y - (\nabla_Y A_{a^*})X = {} & s_a(Y)AX - s_a(X)AY \\
& + \sum_{b=1}^{q}\Big\{s_{a^*b}(X)A_bY - s_{a^*b}(Y)A_bX\Big\} \\
& + \sum_{b=1}^{q}\Big\{s_{a^*b^*}(X)A_{b^*}Y - s_{a^*b^*}(Y)A_{b^*}X\Big\}
\end{aligned}
\tag{15.31}
$$

for $X, Y \in T(M)$ and $a = 1,\ldots,q$.

Finally, we give some examples of CR submanifolds of maximal CR dimension.

*Example* 15.1. From the above discussion, it is clear that real hypersurfaces of a complex manifold are CR submanifolds of maximal CR dimension.    ◇

*Example* 15.2. Let $M'$ be a complex submanifold of $\overline{M}$ with immersion $\imath_1$ and $M$ a real hypersurface of $M'$ with immersion $\imath_0$ and $\imath = \imath_1\imath_0$. We denote by $\xi'$ the unit normal vector field to $M$ in $M'$. Since $\imath_1$ is holomorphic, it follows that $\imath_1 J' = J\imath_1$, where $J'$ is the induced almost complex structure of $M'$ from $J$. Now we have for $X \in T(M)$,

$$J\imath X = J\imath_1\imath_0 X = \imath_1 J'\imath_0 X = \imath_1(\imath_0 F'X + u(X)\xi') = \imath F'X + u(X)\imath_1\xi'.$$

On the other hand, we may write

$$J\imath X = \imath F X + \sum_{a=1}^{p} u^a(X)\xi_a,$$

where $\xi_a(a = 1,\ldots,p)$ are local orthonormal vector fields normal to $M$ in $\overline{M}$. If we choose $\xi$ in such a way that $\xi = \imath_1\xi'$, then $J\imath X = \imath F X + u(X)\xi$. Thus, any real hypersurface $M$ of a complex submanifold $M'$ of $\overline{M}$ is a CR submanifold with maximal CR dimension of $\overline{M}$.    ◇

*Example* 15.3. Let $M'$ be a real hypersurface of $\overline{M}$ and $\imath_1$ be the immersion. Then, for any $X' \in T(M')$, we put

$$J\imath_1 X' = \imath_1 F'X' + u'(X')\xi.$$

Then $F'$ and $u'$ define an almost contact structure of $M'$.

Further, let $M$ be an odd-dimensional $F'$-invariant submanifold of $M'$, that is, such that $F'T(M) \subset T(M)$. Denote by $\imath_0$ the immersion and put $\imath = \imath_1 \circ \imath_0$. We choose a local orthonormal basis of $T^\perp(M)$ in $T(\overline{M})$ in such a way that $\xi_1 = \xi$ and $\xi_2,\ldots,\xi_p$ are orthonormal in $T(M')$. Then, for $X \in T(M)$,

$$J\imath X = \imath F X + \sum_{a=1}^{p} u^a(X)\xi_a.$$

Also we have

$$\begin{aligned}
J\imath X = J\imath_1 \circ \imath_0 X &= \imath_1 F'\imath_0 X + u'(\imath_0 X)\xi \\
&= \imath_1 \circ \imath_0 F X + u'(\imath_0 X)\xi = \imath F X + u'(\imath_0 X)\xi,
\end{aligned}$$

since $M$ is $F'$-invariant submanifold. Comparing the above two equations, we conclude $u^1(X) = u'(\imath_0 X)$, $u^a(X) = 0$, $a = 2,\ldots,p$. Hence, any odd-dimensional $F'$-invariant submanifold of a real hypersurface of $\overline{M}$ is a CR submanifold of maximal CR dimension.    ◇

*Example* 15.4. In Example 15.2, let $M'$ be a totally geodesic complex submanifold of $\overline{M}$. Then, from the Weingarten formula, it follows

$$\overline{\nabla}_X \xi = \overline{\nabla}_X \imath_1 \xi = \imath_1 \nabla'_X \xi' + h'(\imath_0 X, \xi') = \imath_1(-\imath_0 A_0 X) = -\imath A_0 X,$$

where $h'$ and $A_0$ denote the second fundamental form of $M'$ in $\overline{M}$ and the shape operator of $M$ in $M'$, respectively. The last equation implies $D_X \xi = 0$, namely, $\xi$ is parallel with respect to the normal connection $D$.    $\diamondsuit$

*Example* 15.5. Let us assume that, in Example 15.3, the shape operator $A'$ of $M'$ in $\overline{M}$ is of the form

$$A'X' = \lambda X' + \mu u'(X)U', \quad \text{where} \quad U' = \imath_0 U.$$

Consequently,

$$\begin{aligned}
\overline{\nabla}_X \xi_1 = \overline{\nabla}_X \xi &= -\imath_1 A'(\imath_0 X) \\
&= -\imath_1(\lambda \imath_0 X + \mu u'(\imath_0 X)U') = -\imath \lambda X - \mu u^1(X)\imath U.
\end{aligned}$$

This implies that $\xi_1$ is parallel with respect to the normal connection, since $D_X \xi_1 = 0$.    $\diamondsuit$

# 16

# Real hypersurfaces of a complex projective space

Let $M$ be a real hypersurface of a Kähler manifold $(\overline{M}, J)$ and let $\xi$ be its unit normal vector field. Then $M$ is a CR submanifold of maximal CR dimension and $\xi$ is the distinguished normal vector field, used to define the almost contact structure $F$ on $M$, induced from the almost complex structure $J$ of $\overline{M}$. Moreover, since a real hypersurface $M$ of a Kähler manifold $\overline{M}$ has two geometric structures: an almost contact structure $F$ and a submanifold structure represented by the shape operator $A$ with respect to $\xi$, in this section we study the commutativity condition of $A$ and $F$ and we present its geometric meaning.

We begin with several results on complex space forms.

**Theorem 16.1.** *If a real hypersurface $M$ of a complex space form $\overline{M}$ of constant holomorphic sectional curvature $4k$ satisfies $\nabla A = 0$, then the curvature tensor $\overline{R}$ of $\overline{M}$ vanishes identically.*

*Proof.* From (9.21), the Codazzi equation (5.23) becomes

$$k\{g(U, X)FY - g(U, Y)FX - 2g(FX, Y)U\} = (\nabla_X A)Y - (\nabla_Y A)X = 0.$$

Suppose $k \neq 0$ and substitute $U$ instead of $Y$ in the equation above to obtain $FX = 0$ for all $X \in T(M)$. This is a contradiction. Hence $k = 0$ and $\overline{R}$ vanishes identically. □

**Lemma 16.1.** *If the shape operator $A$ of a real hypersurface $M^n$ of a complex space form satisfies $AX = \alpha X$ for any $X \in T(M)$, then $\alpha$ is constant.*

*Proof.* Differentiating $AX = \alpha X$ covariantly, we have $(\nabla_Y A)X = (Y\alpha)X$. By the Codazzi equation, it follows that

$$(X\alpha)Y - (Y\alpha)X = (\nabla_X A)Y - (\nabla_Y A)X$$
$$= k\{g(U, X)FY - g(U, Y)FX - 2g(FX, Y)U\},$$

from which, for an orthonormal basis $\{e_1, \ldots, e_n\}$ of $T_x(M)$,

M. Djorić, M. Okumura, *CR Submanifolds of Complex Projective Space*,
Developments in Mathematics 19, DOI 10.1007/978-1-4419-0434-8_16,
© Springer Science+Business Media, LLC 2010

$$k \sum_{i=1}^{n} \{g(U,X)g(Fe_i, e_i) - g(U, e_i)g(FX, e_i) - 2g(FX, e_i)g(U, e_i)\}$$

$$= (n-1)X\alpha.$$

Since the left-hand members of the above equation vanish identically, it follows $X\alpha = 0$, namely, $\alpha$ is constant. □

Since $\mathbf{P}^n(\mathbf{C})$ is a complex space form with $k = 1$, Theorem 16.1 and Lemma 16.1 imply the following

**Corollary 16.1.** [60] *In $\mathbf{P}^n(\mathbf{C})$ there exists neither totally geodesic real hypersurfaces nor totally umbilical real hypersurfaces.*

Now we give examples of real hypersurfaces of a complex projective space.

*Example* 16.1. [36] Let $\mathbf{S}^{2n+1}$ be a sphere of radius 1 defined by $\sum_{i=0}^{n} z^i \bar{z}^i = 1$ in $\mathbf{C}^{n+1} = \mathbf{C}^{p+1} \oplus \mathbf{C}^{q+1}$, $(p + q = n - 1)$. In $\mathbf{S}^{2n+1}$ we choose two spheres, $\mathbf{S}^{2p+1}$ and $\mathbf{S}^{2q+1}$, in such a way that they lie respectively in complex subspaces $\mathbf{C}^{p+1}$ and $\mathbf{C}^{q+1}$ of $\mathbf{C}^{n+1}$. Then the product $\mathbf{S}^{2p+1} \times \mathbf{S}^{2q+1}$ is a hypersurface of $\mathbf{S}^{2n+1}$ and may be expressed for a fixed $t$ by the following equations:

$$\sum_{i=0}^{p} z^i \bar{z}^i = \cos^2 t, \qquad \sum_{i=p+1}^{n+1} z^i \bar{z}^i = \sin^2 t.$$

Since the $\mathbf{S}^1$ action on $\mathbf{S}^{2p+1} \times \mathbf{S}^{2q+1}$ given by

$$(\theta; z^0, \ldots, z^n) \mapsto (e^{\sqrt{-1}\theta} z^0, \ldots, e^{\sqrt{-1}\theta} z^n) = (w^0, \ldots, w^n)$$

satisfies

$$\sum_{i=0}^{p} w^i \bar{w}^i = \sum_{i=0}^{p} e^{\sqrt{-1}\theta} z^i \overline{e^{\sqrt{-1}\theta} z^i} = \sum_{i=0}^{p} z^i \bar{z}^i = \cos^2 t,$$

$$\sum_{i=p+1}^{n+1} w^i \bar{w}^i = \sum_{i=p+1}^{n+1} e^{\sqrt{-1}\theta} z^i \overline{e^{\sqrt{-1}\theta} z^i} = \sum_{i=p+1}^{n+1} z^i \bar{z}^i = \sin^2 t,$$

the quotient manifold $(\mathbf{S}^{2p+1} \times \mathbf{S}^{2q+1})/\mathbf{S}^1$ is a real hypersurface of $\mathbf{P}^n(\mathbf{C})$. We denote this hypersurface by $M_{p,q}^C$. It is represented by an equivalence class $[(z^0, z^1, \ldots, z^n)]$ containing a point $(z^0, \ldots, z^p, z^{p+1}, \ldots, z^n)$ of $\mathbf{S}^{2p+1} \times \mathbf{S}^{2q+1}$. Note that $M_{p,q}^C$ is congruent to $M_{q,p}^C$.

*Remark* 16.1. Particularly, hypersurface $M_{0,q}^C$ is represented as the equivalence class $[(z^0, z^1, \ldots, z^n)]$ containing a point $(z^0, z^1, \ldots, z^n)$ of $\mathbf{S}^1 \times \mathbf{S}^{2n-1}$ given by

$$z^0 \overline{z}^0 = \cos^2 t, \qquad \sum_{i=1}^{n} z^i \overline{z}^i = \sin^2 t.$$

Then the mapping $f : M^C_{0,n-1} \to \mathbf{S}^{2n-1} \subset \mathbf{C}^n$ defined by

$$f([(z^0, z^1, \ldots, z^n)]) = \left( \frac{z^1}{z^0}, \ldots, \frac{z^n}{z^0} \right)$$

gives a diffeomorphism between $M^C_{0,n-1}$ and $\mathbf{S}^{2n-1}$, since

$$\left| \frac{z^1}{z^0} \right|^2 + \cdots + \left| \frac{z^n}{z^0} \right|^2 = \tan^2 t.$$

Therefore, $M^C_{0,n-1}$ is diffeomorphic to a $(2n-1)$-dimensional sphere. Theorem 19.3 implies that $M^C_{0,n-1}$ (which is congruent to $M^C_{n-1,0}$) is a geodesic hypersphere.                                                                 $\diamond$

*Example* 16.2. [56] We consider in $\mathbf{S}^{2n+1}$ the hypersurface $M'(n+1,t)$ defined by

$$\left| \sum_{j=0}^{l} z_j^2 \right|^2 = t,$$

where $t$ is a fixed positive number $0 < t < 1$. (Note that this is another expression of the hypersurface defined in Example 12.3, using the complex numbers.) The $\mathbf{S}^1$ action to $M'(n+1,t)$ satisfies

$$\left| \sum_{j=0}^{l} w_j^2 \right|^2 = \left| \sum e^{2\sqrt{-1}\theta} \sum_{j=0}^{l} z_j^2 \right|^2 = \left| \sum_{j=0}^{l} z_j^2 \right|^2,$$

and the quotient space $M'(n+1,t)/\mathbf{S}^1$ is a real hypersurface of $\mathbf{P}^n(\mathbf{C})$. We denote it by $M(n,t)$.                                                                 $\diamond$

Let $M$ be a real hypersurface of $\mathbf{P}^n(\mathbf{C})$ and let $\pi^{-1}(M)$ be the circle bundle over $M$ which is compatible with the Hopf map $\pi : \mathbf{S}^{2n+1} \to \mathbf{P}^n(\mathbf{C})$. Then $\pi^{-1}(M)$ is a hypersurface of $\mathbf{S}^{2n+1}$. We denote by $\imath'$ the immersion of $\pi^{-1}(M)$ into $\mathbf{S}^{2n+1}$. The compatibility $\pi \circ \imath' = \imath \circ \pi$ with the Hopf map is expressed by the following commutative diagram:

$$
\begin{array}{ccc}
\pi^{-1}(M) & \xrightarrow{\imath'} & \mathbf{S}^{2n+1} \\
\downarrow{\pi} & & \downarrow{\pi} \\
M & \xrightarrow{\imath} & \mathbf{P}^n(\mathbf{C})
\end{array}
$$

For the unit normal vector field $\xi$ of $M$ to $\mathbf{P}^n(\mathbf{C})$, the horizontal lift $\xi^*$ of $\xi$ is the unit normal vector field of $\pi^{-1}(M)$ to $\mathbf{S}^{2n+1}$. For the vertical vector field $V'$ of the circle bundle $\pi^{-1}(M)(M, \mathbf{S}^1)$, vector field $\imath'V'$ is the vertical vector field of the circle bundle $\mathbf{S}^{2n+1}(\mathbf{P}^n(\mathbf{C}), \mathbf{S}^1)$. As we have shown in Section 9, the integral curve of $\imath'V'$ is a great circle in $\mathbf{S}^{2n+1}$ and therefore the integral curve is a geodesic of $\mathbf{S}^{2n+1}$. Hence we have

$$\nabla^S_{V'}\imath'V' = \imath'\nabla'_{V'}V' + g'(A'V',V')\xi^* = 0$$

and consequently,

$$\nabla'_{V'}V' = 0, \qquad g'(A'V',V') = 0, \tag{16.1}$$

where $A'$ is the shape operator of $\pi^{-1}(M)$.

The condition $\pi \circ \imath' = \imath \circ \pi$ implies that $\imath'X^* = (\imath X)^*$ and using (9.10), we can calculate $\nabla^S_{\imath'X^*}\imath'Y^*$ in the following two ways:

$$\begin{aligned}
\nabla^S_{X^*}\imath'Y^* &= \imath'\nabla'_{X^*}\imath'Y^* + g'(A'X^*,Y^*)\xi^* \\
&= \imath'\{(\nabla_X Y)^* + g(X,FY)V'\} + g'(A'X^*,Y^*)\xi^* \\
&= \imath'(\nabla_X Y)^* + g(X,FY)\imath'V' + g'(A'X^*,Y^*)\xi^*,
\end{aligned}$$

and

$$\begin{aligned}
\nabla^S_{X^*}\imath'Y^* &= \nabla^S_{X^*}(\imath Y)^* = (\overline{\nabla}_X \imath Y)^* - g^S((\imath X)^*, J(\imath Y)^*)\imath'V' \\
&= (\imath \nabla_X Y + g(AX,Y)\xi)^* - g^S(\imath'X^*, \imath'F'Y^*)\imath'V' \\
&= \imath'(\nabla_X Y)^* + g(AX,Y)\xi^* - g'(X^*,F'Y^*)\imath V'.
\end{aligned}$$

Comparing the last two equations we obtain

$$g'(A'X^*,Y^*) = g(AX,Y). \tag{16.2}$$

In the same way,

$$\begin{aligned}
\nabla^S_{X^*}\imath'V' &= \imath'\nabla'_{X^*}V' + g'(A'X^*,V')\xi^*, \\
\nabla^S_{X^*}\imath'V' &= \nabla^S_{X^*}\imath'V' = J' \circ j(\imath X)^* = (J\imath X)^* = (\imath FX + u(X)\xi)^* \\
&= (\imath FX)^* + u(X)\xi^* = \imath'(FX)^* + u(X)\xi^*,
\end{aligned}$$

where $J'$ and $\imath'$ are the natural almost complex structure of $\mathbf{C}^{n+1}$ and the natural immersion of $\mathbf{S}^{2n+1}$ into $\mathbf{C}^n$, respectively, and we used the result of Section 9. Comparing the above two equations, we conclude

$$\nabla'_{X^*}V' = \nabla'_{V'}X^* = (FX)^*, \qquad g'(A'X^*,V') = u(X). \tag{16.3}$$

We note that the first equation in (16.3) follows from the fact that $[V',X^*]$ is vertical. Since $A'$ is symmetric, using the second equation of (16.3), we compute

$$g'(U^*,X^*) = g(U,X) = u(X) = g'(A'X^*,V') = g'(A'V',X^*)$$

and consequently

$$A'V' = U^*.  \tag{16.4}$$

Relations (15.8) and (16.4) imply

$$g'(A'V', A'V') = 1.$$

Using (16.2) and (16.3), we derive the following relation between the shape operators $A$ and $A'$:

$$A'X^* = (AX)^* + g(U, X)V'.  \tag{16.5}$$

**Lemma 16.2.** *For the shape operators $A$ and $A'$ the following relation holds:*

$$\text{trace } A' = \text{trace } A$$

*and therefore $\pi^{-1}(M)$ is minimal if and only if $M$ is minimal.*

*Proof.* Let $E_i$, $i = 1, \ldots, n$ be mutually orthonormal vector fields of $M$. In $T(\pi^{-1}(M))$, we take mutually orthonormal vector fields $E'_i$, $i = 1, \ldots, n+1$, in such a way that $E'_i = E^*_i$, $i = 1, \ldots, n$ and $E'^{n+1} = V'$. Then

$$\text{trace } A' = \sum_{i=1}^{n+1} g'(A'E'_i, E'_i) = \sum_{i=1}^{n} g'(A'E^*_i, E^*_i) + g'(A'V', V').$$

By means of (16.1) and (16.5), it follows

$$\text{trace } A' = \sum_{i=1}^{n} g'((AE_i)^*, E^*_i) - \sum_{i=1}^{n} g'(A'E^*_i, V')g'(V', E^*_i)$$

$$= \sum_{i=1}^{n} g(AE_i, E_i) = \text{trace } A,$$

which completes the proof. $\qquad\square$

**Lemma 16.3.** *Let $\lambda_1, \ldots, \lambda_n$ be the principal curvatures of $M$ and $U$ the eigenvector field corresponding to the principal curvature $\lambda_n$, that is, $AU = \lambda_n U$. Then the principal curvatures of $\pi^{-1}(M)$ are given by $\lambda_1, \ldots, \lambda_{n-1}$, $\mu$ and $-\frac{1}{\mu}$, where $\mu = \frac{\lambda_n \pm \sqrt{\lambda_n + 4}}{2}$.*

*Proof.* Let $X_i$ be the eigenvector field which corresponds to the principal curvature $\lambda_i \neq \lambda_n$. Then, by (16.5),

$$A'X^*_i = (AX_i)^* - g(U, X)V' = (AX)^* = \lambda_i X^*_i.$$

Thus for $i = 1, \ldots, n-1$, $\lambda_i$ are principal curvatures of $\pi^{-1}(M)$. To obtain the other principal curvatures, we put these as $\mu$ and $\nu$. The eigenvectors $X'$, $Y'$ which correspond to $\mu$ and $\nu$ must be linear combinations of $V'$ and $U^*$ and we put

$$X' = V' \cos\theta + U^* \sin\theta, \qquad Y' = -V' \sin\theta + U^* \cos\theta.$$

Then $A'X' = \mu X'$ and $AY' = \nu Y'$ imply that

$$A'V' \cos\theta + A'U^* \sin\theta = \mu V' \cos\theta + \mu U^* \sin\theta,$$
$$-A'V' \sin\theta + A'U^* \cos\theta = -\nu V' \sin\theta + \nu U^* \cos\theta.$$

Comparing the inner product $g'(AX', V')$ and $g'(A'Y', V')$ and making use of (16.1), (16.2) and (16.3), we get $\mu = -\tan\theta$ and $\nu = -\cot\theta$. Thus the principal curvatures of $\pi^{-1}(M)$ are $\lambda_1, \ldots, \lambda_{n-1}, -\tan\theta, \cot\theta$ for some $\theta$. To prove the last part of the Lemma, we recall Lemma 16.2. Consequently, $\lambda_n = \mu + \nu$ and $\mu\nu = -1$, which completes the proof. $\qquad\square$

**Lemma 16.4.** *If the shape operator $A'$ of $\pi^{-1}(M)$ is parallel, then $FA = AF$.*

*Proof.* By (16.2), $g'(A'X^*, Y^*)$ is invariant along the fiber. Using (16.5), it follows

$$
\begin{aligned}
0 &= V'(g'(A'X^*, Y^*)) \\
&= g'((\nabla'_{V'}A')X^*, Y^*) + g'(A'\nabla'_{V'}X^*, Y^*) + g'(A'X^*, \nabla'_{V'}Y^*) \\
&= g'((\nabla'_{V'}A')X^*, Y^*) + g'(AY^*, (FX)^*) + g'(A'X^*, (FY)^*) \\
&= g'((\nabla'_{V'}A')X^*, Y^*) + g(AY, FX) + g(AX, FY),
\end{aligned}
$$

from which, if the shape operator $A'$ is parallel, we conclude $g(FAX, Y) = g(AFX, Y)$, that is, $A$ and $F$ commute. $\qquad\square$

Therefore, $A$ and $F$ commute in the model space $M_{p,q}^C$. Now we consider the converse problem.

First we note that $FA = AF$ implies that $U$ is an eigenvector field of $A$. In fact, $FAU = AFU = 0$ implies that $F^2AU = 0$. This, together with (15.6), implies $AU = \alpha U$, where $\alpha = g(AU, U)$.

Further, differentiating $AU = \alpha U$ covariantly, and using (15.27), we obtain

$$(\nabla_X A)U + AFAX = (X\alpha)U + \alpha FAX.$$

Since $\nabla_X A$ is symmetric $g((\nabla_X A)U, Y) = g((\nabla_X A)Y, U)$ and therefore

$$
\begin{aligned}
g((\nabla_X A)Y &- (\nabla_Y A)X, U) + g(AFAX, Y) - g(AFAY, X) \\
&= (X\alpha)g(U, Y) - (Y\alpha)g(U, X) + \alpha g(FAX, Y) - \alpha g(FAY, X).
\end{aligned}
$$

On the other hand, from the Codazzi equation, it follows

$$
\begin{aligned}
g((\nabla_X A)Y &- (\nabla_Y A)X, U) \\
&= g(U, X)g(FY, U) - g(U, Y)g(FX, U) - 2g(FX, Y)g(U, U) \\
&= -2g(FX, Y).
\end{aligned}
$$

Hence we have

$$-2g(FX,Y) + 2g(AFAX,Y) = (X\alpha)g(U,Y) - (Y\alpha)g(U,X)$$
$$+ 2\alpha g(FAX,Y). \tag{16.6}$$

Substituting $Y$ for $U$ in (16.6) and making use of the fact that $\alpha$ is an eigenvalue of $A$, we obtain $X\alpha = g(U,X)U\alpha$, from which $\operatorname{grad}\alpha = \beta U$ for some $\beta$. Differentiating this covariantly, we have $\nabla_X \operatorname{grad}\alpha = (X\beta)U + \beta FAX$ and therefore

$$0 = g(\nabla_X \operatorname{grad}\alpha, Y) - g(\nabla_Y \operatorname{grad}\alpha, X)$$
$$= (X\beta)g(U,Y) - (Y\beta)g(U,X) + 2\beta g(FX,Y). \tag{16.7}$$

Substituting $Y$ for $U$, we get $X\beta = g(U,X)U\beta$ and then (16.7) becomes $\beta FAX = 0$. Assuming that there exists a point $x \in M$ such that $\beta(x) \neq 0$, it follows $FAX = 0$ at $x$. Therefore, using (16.6), we conclude $g(FX,Y) = 0$ for any $X, Y \in T(M)$. This is a contradiction, since $n > 1$. Hence $\beta = 0$ and $\alpha$ is constant, namely, we proved

**Lemma 16.5.** *If $M$ satisfies the commutative condition $FA = AF$, then $U$ is an eigenvector of $A$ with constant eigenvalue.*

Further, if $FA = AF$, then relation (16.6) becomes

$$F(A^2 X - \alpha AX - X) = 0,$$

since $X\alpha = g(U,X)U\alpha$. Applying $F$ to this equation, we obtain

$$A^2 X - \alpha AX - X + g(U,X)U = 0. \tag{16.8}$$

**Theorem 16.2.** *Let $M$ be a real hypersurface of a complex projective space. The shape operator $A$ and the almost contact tensor $F$ commute if and only if the shape operator $A'$ of $\pi^{-1}(M) \subset S^{2n+1}$ is parallel.*

*Proof.* The necessity is already proved in Lemma 16.4. We now prove the sufficiency. From Theorem 13.2, it is enough to show that if $A$ satisfies (16.8) this implies that $A'$ satisfies (13.5).

For $x \in M$, let $y \in \pi^{-1}(x) \subset \pi^{-1}(M) \subset S^{2n+1}$. Since

$$T_y(\pi^{-1}(M)) = H_y(\pi^{-1}(M)) \oplus \operatorname{span}\{V_y'\},$$

any $X' \in T_y(\pi^{-1}(M))$ can be expressed at $y$ as

$$X_y' = X_y^* + g'(X',V')(y)V_y',$$

where $X$ is a tangent vector at $x$. Hence it is enough to show that $A'$ satisfies (13.5) only for horizontal lift $X^*$ and the unit vertical vector $V'$. Making use of (16.5), we have

$$A'^2 X^* - \alpha A' X^* - X^* = A'((AX)^* + g(U, X)V') - \alpha((AX)^*$$
$$+ g(U, X)V') - X^*$$
$$= (A^2 X - \alpha AX - X + g(U, X)U)^* = 0.$$

Thus for horizontal vector $X'$ at $y$, we obtain

$$A'^2 X' - \alpha A' X' - X' = 0.$$

For a vertical vector $V'$, using (16.4), we have

$$A'^2 V' - \alpha A' V' - V' = A' U^* - \alpha U^* - V'$$
$$= (AU)^* + g(U, U)V' - \alpha U^* - V' = 0.$$

Thus, for any $X' \in T_y(\pi^{-1}(M))$, we have $A'^2 X' - \alpha A' X' - X' = 0$. Hence, by Theorem 11.3, it follows that $A'$ is parallel. $\qquad\square$

Theorems 13.2 and 16.2 imply

**Theorem 16.3.** [45] $M_{p,q}^C$ *is the only complete real hypersurface of a complex projective space whose shape operator $A$ commutes with the almost contact tensor $F$.*

Now we prove a classification theorem of real hypersurface of complex Euclidean space which satisfies the commutative condition $FA = AF$ by almost the same discussion as above. In this case, the Codazzi equation takes the form of (11.5) and in entirely the same way we know that the principal curvature $\alpha$ is constant and the shape operator $A$ satisfies $FA^2 X - \alpha FAX = 0$, from which

$$A^2 X - \alpha X = 0. \tag{16.9}$$

This shows that $M$ has at most two distinct constant principal curvatures and one of them is zero. We consider the case that $M$ has two distinct curvatures $\alpha$ and 0. Let $r$ be the multiplicity of $\alpha$. Note that $AU = \alpha U \neq 0$ and $FAU = \alpha FU = 0$ which means that

$$\text{rank}\,(FA) \le r - 1.$$

Let $X$ be an eigenvector of $A$ corresponding to $\alpha$ which satisfies $FAX = 0$, and $g(X, U) = 0$. Then $F^2 AX = 0$ implies that $AX = g(AX, U) = \alpha g(X, U) = 0$. This means that such a vector $X$ corresponds to the eigenvalue 0. Hence, there exists no other vector than $U$ which satisfies both $FAX = 0$ and $AX = \alpha X$. Thus we get

$$\text{rank}\,(FA) \ge r - 1.$$

Combining the above two inequations, we have $\text{rank}\,(FA) = r - 1$.

On the other hand, putting $\omega(X, Y) = g(FAX, Y)$, we define a 2-form $\omega$, since $FA = AF$. From this fact, it follows $r - 1 = \text{rank}\,(FA) = \text{rank}\,(\omega) = even$. This shows that $r$ is an odd number.

Now, applying Theorem 11.4, we obtain

**Theorem 16.4.** [44] *Let M be a complete real hypersurface of a complex Euclidean space* $\mathbf{E}^{n+1}$. *If M satisfies the commutative condition* $FA = AF$, *then M is one of the following:*

(1) *n-dimensional hypersphere* $\mathbf{S}^n$,

(2) *n-dimensional hyperplane* $\mathbf{E}^n$,

(3) *product manifold of an odd-dimensional sphere and Euclidean space* $\mathbf{S}^r \times \mathbf{E}^{n-r}$.

# Tubes over submanifolds

The examples given in Section 16 are sometimes referred to as tubes over various submanifolds. Therefore in this section we introduce the notion of a tube over a submanifold. For that purpose, let $M$ be a submanifold of a Riemannian manifold $\overline{M}$ and $BM$ the bundle of unit normal vectors of $M$, that is,

$$BM = \bigcup_{x \in M} B_x M$$
$$= \bigcup_{x \in M} \left\{ \xi_x | \xi_x \in T_x^{\perp}(M), |\xi_x| = 1 \right\}.$$

For a sufficiently small real number $t \in \mathbf{R} \setminus \{0\}$, we can define the following immersion:

$$\phi_t : BM \to \overline{M}, \qquad \phi_t(\xi) = \exp t\xi,$$

where exp denotes the exponential mapping of $\overline{M}$. This $\phi_t(BM)$ with induced Riemannian metric from $\overline{M}$ is called the *tube of radius $t$ over $M$ in $\overline{M}$*.

We illustrate this notion of a tube, beginning with Examples 17.1–17.4, which are elementary and well-known examples.

*Example* 17.1. Let $x$ be a point of $\overline{M}$. Then

$$B_x M = \left\{ X_x \,|\, X_x \in T_x(\overline{M}), |X_x| = 1 \right\}$$

and $\phi_t(BM)$ is the focus of all points whose geodesic distance from $x$ is $t$. Thus, the tube over the 0-dimensional manifold $x$, namely, over one point, is a geodesic hypersphere centered at $x$.  $\diamondsuit$

*Example* 17.2. Let $\mathbf{S}^1(1)$ denote a circle in $\mathbf{E}^3$ defined by

$$\mathbf{S}^1(1) = \left\{ \mathbf{x} = (\cos u, \sin u, 0) \in \mathbf{E}^3 \right\}$$

and let $\mathbf{x}$ denote the position vector field of $\mathbf{S}^1(1)$. Then

M. Djorić, M. Okumura, *CR Submanifolds of Complex Projective Space*,
Developments in Mathematics 19, DOI 10.1007/978-1-4419-0434-8_17,
© Springer Science+Business Media, LLC 2010

$$\mathbf{n}_1 = (-\cos u, -\sin u, 0),$$
$$\mathbf{n}_2 = (0, 0, 1)$$

are mutually orthonormal unit normal vectors to $\mathbf{S}^1(1)$ in $\mathbf{E}^3$ and any unit normal vector $\xi$ to $\mathbf{S}^1(1)$ in $\mathbf{E}^3$ is given by

$$\xi = (-\cos v \cos u, -\cos v \sin u, \sin v).$$

Since the geodesic in $\mathbf{E}^3$ is a straight line and the position vector $\mathbf{y}$ of $\phi_t(\mathbf{S}^1(1))$ is given by

$$\mathbf{y} = \mathbf{x} + t\xi = (\cos u(1 - t\cos v), \sin u(1 - t\cos v), t\sin v),$$

we conclude that the tube over $\mathbf{S}^1(1)$ in $\mathbf{E}^3$ is a torus.    $\Diamond$

*Example* 17.3. Now we consider the circle in Example 17.2 as a special curve of $\mathbf{S}^2(1)$. Then, $BM = \{(0,0,1), (0,0,-1)\}$. Since the geodesic $\gamma(t)$ of $\mathbf{S}^2(1)$ in the direction of $\xi$ is represented by

$$\gamma(t) = \cos t\mathbf{x} + \sin t\xi,$$

where $\mathbf{x}$ is the position vector of $\mathbf{S}^1(1)$, it follows

$$\phi_t(\xi) = (\cos t \cos u, \cos t \sin u, \pm \sin t).$$

Thus, the tube over $\mathbf{S}^1(1)$ in $\mathbf{S}^2(1)$ is the union of two circles of radius $|\cos t|$ near the original circle $\mathbf{S}^1(1)$.    $\Diamond$

*Example* 17.4. Let $\mathbf{S}^n(1)$ be the sphere which is the totally geodesic submanifold of $\mathbf{S}^{n+p}(1)$ defined by $(\mathbf{x}, \mathbf{0})$, where $\mathbf{x}$ denotes the position vector of a point in $\mathbf{S}^n(1)$ in $\mathbf{E}^{n+1}$ and $\mathbf{0} = (0, \dots, 0)$ denotes the zero vector in the $p$-dimensional Euclidean space $\mathbf{E}^p$. Identifying $T_{\mathbf{x}}\mathbf{E}^{n+p+1}$ with $\mathbf{E}^{n+p+1}$, the set of unit normal vectors to $\mathbf{S}^n(1)$ in $\mathbf{S}^{n+p}(1)$ at $\mathbf{x}$ is

$$B_{\mathbf{x}}\mathbf{S}^n(1) = \{(\mathbf{0}, \mathbf{y}) \mid |\mathbf{y}| = 1\}, \ \mathbf{0} \in \mathbf{E}^{n+1}, \ \mathbf{y} \in \mathbf{E}^p.$$

Since the geodesic $\gamma(t)$ of $\mathbf{S}^{n+p}(1)$ in the direction of $\xi_{\mathbf{x}}$ is $\gamma(t) = \cos t\mathbf{x} + \sin t\xi_{\mathbf{x}}$, we have

$$\phi_t(BS^n(1)) = \{(\cos t\mathbf{x}, \sin t\mathbf{y}) \mid |\mathbf{x}| = |\mathbf{y}| = 1\}$$
$$= \mathbf{S}^n(|\cos t|) \times \mathbf{S}^{p-1}(|\sin t|).$$    $\Diamond$

Now we prove that real hypersurfaces $M_{n,m}^C$ and $M(n,t)$, introduced in Section 16, are tubes over some submanifolds.

Let $\mathbf{P}^{\frac{n}{2}}(\mathbf{C})$ be a totally geodesic, complex projective subspace of $\mathbf{P}^{\frac{n+p}{2}}(\mathbf{C})$ and let $\pi : \mathbf{S}^{n+p+1}(1) \to \mathbf{P}^{\frac{n+p}{2}}(\mathbf{C})$ be the Hopf map. We deduce from Proposition 10.2 that $\pi^{-1}(\mathbf{P}^{\frac{n}{2}}(\mathbf{C})) = \mathbf{S}^{n+1}(1)$ is a totally geodesic submanifold of

$\mathbf{S}^{n+p+1}(1)$. For a unit normal vector $\xi_x$ at $x \in \mathbf{P}^{\frac{n}{2}}(\mathbf{C})$, the exponential map $\phi_t : B_x\mathbf{P}^{\frac{n}{2}}(\mathbf{C}) \to \mathbf{P}^{\frac{n+p}{2}}(\mathbf{C})$ is given by

$$\phi_t(\xi_x) = \pi(\cos tw + \sin t\xi_w') = \pi\phi_t'(\xi_w'),$$

where $w$ is a point of $\pi^{-1}(\mathbf{P}^{\frac{n}{2}}(\mathbf{C})) = \mathbf{S}^{n+1}(1)$ such that $\pi(w) = x$. Here $\xi_w'$ denotes the horizontal lift of $\xi_x$ at $w$ and $\phi_t'$ denotes the exponential map $B_w\mathbf{S}^{n+1}(1) \to \mathbf{S}^{n+p+1}(1)$. This shows that any point of the tube over $\mathbf{P}^{\frac{n}{2}}(\mathbf{C})$ is a $\pi$-image of a point of the tube around $\mathbf{S}^{n+1}(1)$. This, together with Example 17.4, implies that the tube around $\mathbf{P}^{\frac{n}{2}}(\mathbf{C})$ is

$$B\mathbf{P}^{\frac{n}{2}}(\mathbf{C}) = \pi(B\mathbf{S}^{n+1}(1)) = \pi(\mathbf{S}^{n+1}(|\cos t|) \times \mathbf{S}^{p-1}(|\sin t|)) = M^C_{n_1, n_2},$$

where $n_1 = \frac{n}{2}$ and $n_2 = \frac{p-1}{2}$. Thus, we have

**Proposition 17.1.** $M^C_{n,m}$ *is a tube over the totally geodesic complex subspace* $\mathbf{P}^{\frac{n}{2}}(\mathbf{C})$ *in* $\mathbf{P}^{\frac{n+p}{2}}(\mathbf{C})$.

Particularly, for the case $p = 1$, using Remark 16.1, it follows $B\mathbf{P}^{\frac{n}{2}}(\mathbf{C}) = M^C_{n_1,0}$, which leads to

**Corollary 17.1.** *The geodesic hypersphere* $M^C_{n_1,0}$ *is a tube over totally geodesic complex hyperplane.*

Now we consider the tube over a real projective space in a complex projective space. Let $(u_1, v_1, \ldots, u_{n+1}, v_{n+1})$ be homogeneous coordinates of $\mathbf{P}^n(\mathbf{C})$ in $\mathbf{C}^{n+1} = \mathbf{R}^{2n+2}$, that is,

$$(u_1, v_1, \ldots, u_{n+1}, v_{n+1}) \in \mathbf{S}^{2n+1}, \quad \sum_{i=1}^{n+1}(u_i^2 + v_i^2) = 1.$$

Then, as a submanifold of $\mathbf{P}^n(\mathbf{C})$, a real projective space $\mathbf{P}^n(\mathbf{R})$ is represented by

$$(u_1, 0, \ldots, u_{n+1}, 0), \quad \sum_{i=1}^{n+1} u_i^2 = 1.$$

$(u_1, 0, \ldots, u_{n+1}, 0)$ belongs to the equivalence class

$$[(u_1\cos\theta, u_1\sin\theta, \ldots, u_{n+1}\cos\theta, u_{n+1}\sin\theta)], \quad \sum_{i=1}^{n+1} u_i^2 = 1$$

in $\mathbf{S}^{2n+1}$. That is, the position vector field $\mathbf{w}$ of the submanifold

$$M^{n+1} = \pi^{-1}(\mathbf{P}^n(\mathbf{R})) \subset \mathbf{S}^{2n+1} \subset \mathbf{C}^{n+1} = \mathbf{R}^{2n+2}$$

is given by

$$\mathbf{w} = \sum_{i=1}^{n+1} u_i \left( \cos\theta \frac{\partial}{\partial u_i} + \sin\theta \frac{\partial}{\partial v_i} \right),$$

where $\pi$ is the Hopf fibration $\pi : \mathbf{S}^{2n+1} \to \mathbf{P}^n(\mathbf{C})$. Since the tangent space of $M^{n+1}$ is given by

$$T_{\mathbf{w}}(M^{n+1}) = \text{span} \left\{ \cos\theta \frac{\partial}{\partial u_i} + \sin\theta \frac{\partial}{\partial v_i} \right\}, \quad i = 1, \ldots, n+1,$$

the unit normal $\xi$ to $M^{n+1}$ in $\mathbf{S}^{2n+1}$ is

$$\xi = \sum_{j=1}^{n+1} a_j \left( -\sin\theta \frac{\partial}{\partial u_j} + \cos\theta \frac{\partial}{\partial v_j} \right),$$

where $\sum_{j=1}^{n+1} a_j^2 = 1$. For the natural almost complex structure $J$ of $\mathbf{C}^{n+1}$, it follows that

$$J\mathbf{w} = \sum_{i=1}^{n+1} u_i \left( -\sin\theta \frac{\partial}{\partial u_i} + \cos\theta \frac{\partial}{\partial v_i} \right)$$

is the vertical vector field with respect to $\pi$ and, therefore, tangent to $M^{n+1}$. Hence we have

$$\langle J\mathbf{w}, \xi \rangle = \sum_{i=1}^{n+1} a_i u_i = 0. \tag{17.1}$$

The position vector field $\mathbf{z}$ of the tube of radius $t$ over $M^{n+1} \subset \mathbf{S}^{2n+1}$ is

$$\mathbf{z} = \mathbf{w} \cos t + \xi \sin t$$
$$= \sum_{i=1}^{n+1} \left\{ (u_i \cos\theta \cos t - a_i \sin\theta \sin t) \frac{\partial}{\partial u_i} + (u_i \sin\theta \cos t + a_i \cos\theta \sin t) \frac{\partial}{\partial v_i} \right\}.$$

We put

$$x_i = u_i \cos\theta \cos t - a_i \sin\theta \sin t,$$
$$y_i = u_i \sin\theta \cos t + a_i \cos\theta \sin t.$$

Then $(x_1, y_1, \ldots, x_{n+1}, y_{n+1})$ are coordinates of the tube over $M^{n+1} \subset \mathbf{S}^{2n+1} \subset \mathbf{R}^{2n+2}$. It is an easy matter to use (17.1) and obtain

$$\left\{ \sum_{i=1}^{n+1} (x_i^2 - y_i^2) \right\}^2 + 4 \left( \sum_{i=1}^{n+1} x_i y_i \right)^2 = \cos^2 2t,$$

which is the equation from Example 12.3. Using the arguments of Example 17.2 and Proposition 17.1, we have

**Proposition 17.2.** *$M(n,t)$ is a tube of radius $t$ over the real projective space $\mathbf{P}^n(\mathbf{R})$ in $\mathbf{P}^n(\mathbf{C})$.*

Next, we show that $M(n,t)$ is also a tube over a complex quadric $Q^{n-1}$ in a complex projective space $\mathbf{P}^n(\mathbf{C})$. A *complex quadric* $Q^{n-1}$ is defined by

$$Q^{n-1} = \left\{ (w_1, \ldots, w_{n+1}) \in \mathbf{P}^n(\mathbf{C}) \mid \sum_{i=1}^{n+1} w_i^2 = 0 \right\},$$

where $w_i$, $i = 1, \ldots, n+1$ are homogeneous coordinates of $\mathbf{P}^n(\mathbf{C})$. Thus $Q^{n-1}$ can be equivalently defined as

$$Q^{n-1} = \left\{ \pi \left( \frac{1}{\sqrt{2}} (u + \sqrt{-1}v) \right) \mid u \in \mathbf{S}^n, \, v \in \mathbf{S}^n \subset \mathbf{R}^{n+1}, \langle u, v \rangle = 0 \right\},$$

where $\pi$ is the Hopf fibration $\pi : \mathbf{S}^{2n+1} \to \mathbf{P}^n(\mathbf{C})$, that is,

$$\langle u, v \rangle = \sum_{i=1}^{n+1} u_i v_i = 0, \tag{17.2}$$

$$(u_1, \ldots, u_{n+1}) \in \mathbf{S}^n \subset \mathbf{R}^{n+1}, \, \sum_{i=1}^{n+1} u_i^2 = 1, \tag{17.3}$$

$$(v_1, \ldots, v_{n+1}) \in \mathbf{S}^n \subset \mathbf{R}^{n+1}, \, \sum_{i=1}^{n+1} v_i^2 = 1. \tag{17.4}$$

Let $M^{2n-1} = \pi^{-1}(Q^{n-1}) \subset \mathbf{S}^{2n+1}$. Then we can express $M^{2n-1}$ by

$$M^{2n-1} = \left\{ \frac{1}{\sqrt{2}} (u_1, v_1, \ldots, u_{n+1}, v_{n+1}) \in \mathbf{S}^{2n+1} \subset \mathbf{R}^{2n+2} \right\}.$$

Since the normal space $T^\perp(M^{2n-1})$ in $\mathbf{S}^{2n+1}$ is spanned by the following two unit vectors:

$$\eta = \frac{1}{\sqrt{2}} \sum_{i=1}^{n+1} \left( u_i \frac{\partial}{\partial u_i} - v_i \frac{\partial}{\partial v_i} \right),$$

$$J\eta = \frac{1}{\sqrt{2}} \sum_{i=1}^{n+1} \left( v_i \frac{\partial}{\partial u_i} + u_i \frac{\partial}{\partial v_i} \right),$$

any unit normal vector $\xi$ to $M^{2n-1}$ in $\mathbf{S}^{2n+1}$ is given by

$$\begin{aligned} \xi &= \eta \cos\theta + J\eta \sin\theta \\ &= \frac{1}{\sqrt{2}} \sum_{i=1}^{n+1} \left\{ (u_i \cos\theta + v_i \sin\theta) \frac{\partial}{\partial u_i} + (-v_i \cos\theta + u_i \sin\theta) \frac{\partial}{\partial v_i} \right\}. \end{aligned}$$

Thus, for the position vector field $\mathbf{w}$ of $M^{2n-1}$, the position vector $\mathbf{z}$ of a tube $BM^{2n-1}$ is given by

$$\mathbf{z} = \mathbf{w}\cos t + \xi\sin t$$

$$= \frac{1}{\sqrt{2}}\sum_{i=1}^{n+1}\left\{[u_i(\cos t + \cos\theta\sin t) + v_i\sin\theta\sin t]\frac{\partial}{\partial u_i}\right.$$

$$\left. + [v_i(\cos t - \cos\theta\sin t) + u_i\sin\theta\sin t]\frac{\partial}{\partial v_i}\right\}.$$

Setting

$$x_i = \frac{1}{\sqrt{2}}\left\{u_i(\cos t + \cos\theta\sin t) + v_i\sin\theta\sin t\right\},$$

$$y_i = \frac{1}{\sqrt{2}}\left\{v_i(\cos t - \cos\theta\sin t) + u_i\sin\theta\sin t\right\},$$

$(x_1, y_1, \ldots, x_{n+1}, y_{n+1})$ define coordinates of the tube $B_{\mathbf{w}}M^{2n-1}$. By straight-forward computation, using (17.2), (17.3) and (17.4), we obtain

$$\left\{\sum_{i=1}^{n+1}(x_i^2 - y_i^2)\right\}^2 + 4\left(\sum_{i=1}^{n+1}x_iy_i\right)^2 = \sin^2 2t = \cos^2 2\left(\frac{\pi}{4} - t\right).$$

Thus, we have

**Proposition 17.3.** $M(n, t)$ is a tube of radius $\frac{\pi}{4} - t$ over the complex quadric $Q^{n-1}$ in $\mathbf{P}^n(\mathbf{C})$.

*Remark* 17.1. Using the notation of Example 12.3, we conclude

- $\mathbf{P}^n(\mathbf{R})$ satisfies $F(x, y) = 1$, that is, $\theta = 0$;
- $Q^{n-1}$ satisfies $F(x, y) = 0$, that is, $\theta = \frac{\pi}{4}$.

# Levi form of CR submanifolds of maximal CR dimension of a complex space form

Considering the Levi form on CR submanifolds $M^n$ of maximal CR dimension of complex space forms $\overline{M}^{\frac{n+p}{2}}$, we prove in this section that on some remarkable real submanifolds of complex projective space the Levi form can never vanish and we determine all such submanifolds in the case when the ambient manifold is a complex Euclidean space.

In the following, we establish several formulas in the case when $U$ is the eigenvector of the shape operator $A$. Let $U$ be an eigenvector of $A$ corresponding to the eigenvalue $\alpha$. Taking the covariant derivative of $AU = \alpha U$ and using (15.27), we obtain

$$(\nabla_X A)U + AFAX = (X\alpha)U + \alpha FAX$$

and hence

$$g((\nabla_X A)Y, U) + g(AFAX, Y) = (X\alpha)g(U, Y) + \alpha g(FAX, Y).$$

Thus

$$g((\nabla_X A)Y - (\nabla_Y A)X, U) + 2g(AFAX, Y) = \\ (X\alpha)g(U, Y) - (Y\alpha)g(U, X) + 2\alpha g(FAX, Y).$$

Consequently the Codazzi equation (15.29) yields

$$2k(FY, X) + 2g(AFAX, Y) \tag{18.1}$$
$$= (X\alpha)u(Y) - (Y\alpha)u(X) + \alpha g((FA + AF)X, Y).$$

Putting $Y = U$ in (18.1) and making use of $AY = \alpha U$, we get

$$X\alpha = u(X)U\alpha.$$

This, together with (18.1), implies that

$$-2kFX + 2AFAX = \alpha(FA + AF)X. \tag{18.2}$$

M. Djorić, M. Okumura, *CR Submanifolds of Complex Projective Space*,
Developments in Mathematics 19, DOI 10.1007/978-1-4419-0434-8_18,
© Springer Science+Business Media, LLC 2010

**Lemma 18.1.** *Let $U$ be an eigenvector of $A$ corresponding to the eigenvalue $\alpha$ and let $X$ be the eigenvector of $A$ corresponding to the second eigenvalue $\lambda$. Then we have*

$$(2\lambda - \alpha)AFX = (2k + \alpha\lambda)FX.$$

*Proof.* Let $X$ be an eigenvector of $A$ which corresponds to $\lambda$, then, from (18.2), it follows

$$-2kFX + 2\lambda AFX = \alpha\lambda FX + \alpha AFX,$$

from which Lemma 18.1 follows.                                         □

Now, let us consider the Levi form of CR submanifolds of maximal CR dimension of a complex space form. Using Theorem 8.1, we have

$$L(X,Y) = \sum_{a=1}^{p}\{g(A_aX,Y) + g(A_aFX,FY)\}\xi_a, \tag{18.3}$$

for $X, Y \in H(M)$. Next, we note that for any

$$X, Y \in H(M) = T(M) \cap JT(M),$$

there exist $V, W \in T(M)$ such that

$$\imath X = J\imath V = \imath FV, \quad \imath Y = J\imath W = \imath FW \quad \text{and} \quad FV, FW \in H(M).$$

Using this notation, we may write (18.3) as follows:

$$L(X,Y) = \sum_{a=1}^{p}\{g(A_aFV, FW) + g(A_aF^2V, F^2W)\}\xi_a$$

$$= \sum_{a=1}^{p}\{g(A_aFV, FW) + g(A_aV, W) - u(V)g(A_aU, W)$$

$$- u(W)g(A_aU, V) + u(V)u(W)g(A_aU, U)\}\xi_a, \tag{18.4}$$

where we have used (15.2) and (15.6).

We assume now that $U$ is an eigenvector of $A$, corresponding to the eigenvalue $\alpha$ and that the Levi form vanishes at a point $x \in M$. Using relation (18.4) for $a = 1$, we obtain

$$g(AV - FAFV - \alpha u(V)U, W) = 0. \tag{18.5}$$

We note that here $W$ is chosen in such a way that $Y = FW$, $Y \in H(M)$. However, since

$$g(AV - FAFV - \alpha u(V)U, U) = 0, \tag{18.6}$$

using (18.5) and (18.6), we get

$$AV - FAFV - \alpha u(V)U = 0. \tag{18.7}$$

Now, let $V_x$ be an eigenvector of $A$ at $x$ with the eigenvalue $\lambda$, such that $V_x$ is orthogonal to $U$ at $x$. Then Lemma 18.1 and relation (18.7) imply

$$k + \lambda^2 = 0. \tag{18.8}$$

Thus we have

**Theorem 18.1.** *Let $M$ be an $n$-dimensional $(n \geq 3)$ CR submanifold of CR dimension $\frac{n-1}{2}$ of a complex space form. If $U$ is an eigenvector of the shape operator $A$ with respect to $\xi$, then the Levi form vanishes only when the holomorphic sectional curvature of the ambient manifold is nonpositive.*

**Definition 18.1.** A real hypersurface $M$ is called *strictly pseudoconvex* if, at each point of $M$, the Levi form is either positive or negative definite. If the Levi form is semi-definite, then $M$ is called *pseudoconvex*.

**Theorem 18.2.** *[19] Let $M$ be an $n$-dimensional $(n \geq 3)$ CR submanifold of CR dimension $\frac{n-1}{2}$ of a complex projective space. If $M$ satisfies the conditions of Theorem 18.1, then the Levi form cannot vanish identically. Especially, if $M$ is a real hypersurface, then $M$ is pseudoconvex.*

**Corollary 18.1.** *In $M_{p,q}^C$, the Levi form can never vanish, that is, $M_{p,q}^C$ is pseudoconvex.*

*Proof.* The construction of $M_{p,q}^C$ and Lemma 16.2 imply that $FA = AF$, from which we easily see that $U$ is an eigenvector of $A$. Thus from Theorem 18.2, the corollary follows. □

Now, let the ambient space $\overline{M}$ be an even-dimensional Euclidean space $\mathbf{E}^{n+p}$ equipped with its natural Kaehler structure, that is, $\overline{M} = \mathbf{C}^{(n+p)/2}$. In this case, using the relation (18.8), it follows that all eigenvalues of the shape operator $A$, except the one corresponding to $U$, are equal to zero and therefore the shape operator $A$ can be diagonalized as follows:

$$A = \begin{pmatrix} \alpha & & & \\ & 0 & & \\ & & \ddots & \\ & & & 0 \end{pmatrix}.$$

In the remainder of this section we assume that $\xi = \xi_1$ is parallel with respect to the normal connection. Then in the case of $\alpha = 0$, Theorem 14.1 implies that there exists a totally geodesic hypersurface $\mathbf{E}^{n+p-1}$ of $\mathbf{C}^{\frac{n+p}{2}}$ which contains $M$.

Further, let $\alpha \neq 0$. Then $\alpha$ is not necessarily constant and it may take the value zero at some point. However, $U$ is an eigenvector which never vanishes since it has unit length. This implies that $AX = 0$ for any $X$ orthogonal to $U$.

Now, let $\mathcal{D}$ be the distribution determined by the tangent vectors orthogonal to $U$. Then it follows from relation (15.27) that $g([X,Y],U) = 0$ for all $X, Y \in \mathcal{D}$ and hence, the distribution $\mathcal{D}$ is involutive. Moreover, relation (15.27) implies $\nabla_X U = 0$ for all $X$ tangent to $M$ and we conclude that $M$ is locally a product of $M_\mathcal{D}$ and a curve tangent to $U$, where $M_\mathcal{D}$ denotes an integral submanifold of $\mathcal{D}$. Also, from $\nabla_X U = 0$, we derive that $M_\mathcal{D}$ is a totally geodesic hypersurface of $M$ and consequently the shape operator of $M_\mathcal{D}$ for the normal $U$ vanishes identically. Furthermore, since $A = 0$ on $\mathcal{D}$, it follows that the first normal space of $M_\mathcal{D}$ is a subspace of span $\{\xi_2, \dots, \xi_p\}$.

Further, as $\xi = \xi_1$ is parallel with respect to the normal connection, it follows

$$\overline{g}(D_X \xi_a, \xi) = -\overline{g}(\xi_a, D_X \xi) = 0, \quad \text{for} \quad a = 2, \dots, p.$$

Moreover, using the second relation in (5.8), it follows $\overline{g}(D_X \xi_a, U) = 0$ for $a = 2, \dots, p$. Therefore, span $\{\xi_2, \dots, \xi_p\}$ is invariant under the parallel translation with respect to the normal connection of $M_\mathcal{D}$ in $\mathbf{C}^{\frac{n+p}{2}}$ and Theorem 14.1 then implies that there exists a totally geodesic $\mathbf{E}^{n+p-2}$ in $\mathbf{C}^{\frac{n+p}{2}}$ which contains $M_\mathcal{D}$.

From these results follows

**Theorem 18.3.** [19] *Let $M$ be an $n$-dimensional $(n \geq 3)$ CR submanifold of CR dimension $\frac{n-1}{2}$ of $\mathbf{C}^{\frac{n+p}{2}}$ such that the distinguished normal vector field $\xi$ to $M$ is parallel with respect to the normal connection and $U$ is an eigenvector of the shape operator $A$ with respect to $\xi$. Then the Levi form can vanish only in the following two cases:*

(1) *$M$ is contained in a hyperplane orthogonal to $\xi$;*

(2) *$M$ is locally a Riemannian product $\gamma \times M_\mathcal{D}$, where $\gamma$ is a curve tangent to $U$ and $M_\mathcal{D}$ is contained in an $(n+p-2)$-dimensional subspace $\mathbf{E}^{n+p-2}$.*

*In the second case $M$ is a CR-product, that is, it is locally a product of a holomorphic submanifold $M_\mathcal{D}$ and a totally real submanifold $\gamma$.*

# Eigenvalues of the shape operator $A$ of CR submanifolds of maximal CR dimension of a complex space form

In this section, we assume that $M$ is an $n(\geq 3)$-dimensional CR submanifold of maximal CR dimension of a complex space form $\overline{M}$ with constant holomorphic sectional curvature $4k$ and that the distinguished normal $\xi$ is parallel with respect to the normal connection. Then from Lemma 15.2, the Codazzi equation becomes

$$(\nabla_X A)Y - (\nabla_Y A)X = k\left\{u(X)FY - u(Y)FX - 2g(FX,Y)U\right\}. \quad (19.1)$$

We first prove

**Theorem 19.1.** *If the shape operator $A$ for $\xi$ has only one eigenvalue, then $\overline{M}$ is a complex Euclidean space.*

*Proof.* According to the assumption, it follows that $A = 0$ or $AX = \alpha X$ for all $X \in T(M)$. In both cases the Codazzi equation (19.1) implies

$$(X\alpha)Y - (Y\alpha)X = k\left\{u(X)FY - u(Y)FX - 2g(FX,Y)U\right\},$$

for all $X, Y \in T(M)$. Putting $Y = U$, the last equation reduces to

$$(U\alpha)X - (X\alpha)U = kFX.$$

Since $\dim M \geq 3$, we can choose $U$, $X$ and $FX$ in such a way that they are linearly independent and, hence, $k = 0$. This completes the proof. $\square$

From now on we suppose that the dimension of the submanifold $M$ is greater than three. Further, we assume that the shape operator $A$ has exactly two distinct eigenvalues: $\lambda$ and $\mu$. We are going to prove that, in this case, one of the eigenvectors must be $U$. To that purpose, we denote by $T_\lambda$ and $T_\mu$ the eigenspaces corresponding to the eigenvalue $\lambda$ and $\mu$, respectively. Now suppose

$$U = pX + qV, \quad (19.2)$$

M. Djorić, M. Okumura, *CR Submanifolds of Complex Projective Space*,
Developments in Mathematics 19, DOI 10.1007/978-1-4419-0434-8_19,
© Springer Science+Business Media, LLC 2010

for nonzero functions $p$ and $q$, where $X \in T_\lambda$ and $V \in T_\mu$ are unit vector fields on some open subset of $M$ where $U$ is not an eigenvector of the shape operator $A$. Since dim $M > 3$, at least one of $T_\lambda$ and $T_\mu$ has dimension minimum three, and we may suppose that $\dim T_\lambda \geq 3$. Then, denoting

$$S_\lambda = \{Y \in T_\lambda | g(Y, X) = 0\}, \qquad S_\mu = \{W \in T_\mu | g(W, V) = 0\},$$

we can choose $Y$, $Z$ mutually orthonormal in $S_\lambda$, since $\dim S_\lambda \geq 2$. Then, from (19.2) it follows that $Y$ and $Z$ are orthogonal to $U$ and the Codazzi equation (19.1) becomes

$$(Z\lambda)Y - (Y\lambda)Z + (\lambda I - A)[Z, Y] = 2kg(FY, Z)U. \tag{19.3}$$

Since $U$ and $(\lambda I - A)[Z, Y]$ are orthogonal to $Y$ and $Z$, we conclude that $Y\lambda = 0$ for all $Y \in S_\lambda$. Now, relation (19.3) reduces to

$$(\lambda I - A)[Z, Y] = 2 \, k \, g(FY, Z)U$$

where the left-hand side is orthogonal to $T_\lambda$. Therefore, if $k \neq 0$, $FY$ is orthogonal to $S_\lambda$ for every $Y \in S_\lambda$, since $p$ and $q$ are nonzero functions.

Further, we consider the Codazzi equation for $Y \in S_\lambda$ and the particular vector field $X$. Since $X$, $Y$ are in $T_\lambda$, it reduces to

$$(X\lambda)Y - (Y\lambda)X + (\lambda I - A)[X, Y] = k\{pFY - 2g(FX, Y)U\}. \tag{19.4}$$

We have shown that $Y\lambda = 0$, and we now conclude that $X\lambda = 0$ for all $X \in T_\lambda$, since $(\lambda I - A)[X, Y]$, $FY$ and $U$ are orthogonal to $Y$. Thus, relation (19.4) becomes

$$(\lambda I - A)[X, Y] = k \, \{pFY - 2g(FX, Y)U\}. \tag{19.5}$$

Now, the left-hand side of relation (19.5) is orthogonal to $X \in T_\lambda$ and we obtain

$$3 \, k \, p \, g(FX, Y) = 0.$$

Therefore $FY$ is orthogonal to $X$, because $k \neq 0$ and $p$ is a nonzero function. As a result, $FY \in T_\mu$, since we proved that $FY$ is orthogonal to $S_\lambda$ and to $X$. Moreover,

$$0 = g(FY, U) = qg(FY, V),$$

which means that $FY$ is orthogonal to $V$, because $q \neq 0$. We have thus shown that $FY \in S_\mu$, that is, $F(S_\lambda) \subset S_\mu$, since we proved that $FY \in T_\mu$ and $g(FY, V) = 0$.

Finally from the assumption (19.2) it follows that all $Y \in S_\lambda$ are orthogonal to $U$ and then, according to (15.2), we have $\imath FY = J\imath Y$, for all $Y \in S_\lambda$. This means that $F$ is injective on $S_\lambda$ and thus, $\dim S_\mu \geq \dim S_\lambda \geq 2$. Consequently, we may reverse the roles of the above-specified $\lambda$ and $\mu$ to show that

$F(S_\mu) \subset S_\lambda$. Hence, $\dim S_\mu = \dim S_\lambda$ and therefore, $M$ is even-dimensional, since, taking into account $X$ and $V$, we have $\dim M = 2\dim S_\lambda + 2$. This is a contradiction, and we conclude that $p$ or $q$ is identically zero, and $U$ is an eigenvector of the shape operator $A$. Thus the following lemma holds:

**Lemma 19.1.** *Let $\overline{M}$ be an $(n+p)$-dimensional Kähler manifold of constant holomorphic sectional curvature $4k$, with $n > 3$ and $k \neq 0$. Then, assuming that the shape operator $A$ has exactly two distinct eigenvalues, it follows that $U$ is an eigenvector of $A$.*

**Lemma 19.2.** *If $AF + FA = 0$ holds at a point of the submanifold $M$, then the holomorphic sectional curvature of the ambient manifold is nonpositive.*

*Proof.* It follows from the assumption of the lemma and relation (15.7) that $U$ is an eigenvector of the shape operator $A$, that is, $AU = \alpha U$. Differentiating this equation covariantly and making use of the Codazzi equation (19.1), we obtain

$$2kg(FX, Y) + 2g(AFAX, Y) = (X\alpha)u(Y) - (Y\alpha)u(X)$$
$$+ \alpha g((FA + AF)X, Y). \qquad (19.6)$$

Putting $Y = U$ in relation (19.6), we get

$$X\alpha = u(X)U\alpha, \qquad (19.7)$$

since $U$ is an eigenvector of the shape operator $A$. Using relations (19.6) and (19.7), it follows

$$2kg(FX, Y) + 2g(AFAX, Y) = \alpha g((FA + AF)X, Y), \qquad (19.8)$$

from which $kg(FY, X) = g(AX, FAY)$. Putting $X = FY$ in the last relation, we obtain

$$kg(FY, FY) = g(AFY, FAY) = -g(FAY, FAY) \leq 0,$$

and therefore $k \leq 0$, since $\mathrm{rank}\, F = n - 1$. $\qquad \square$

Until further notice we assume that $\overline{M}$ is a Kähler manifold of constant positive holomorphic sectional curvature.

**Lemma 19.3.** *Let $U$ be an eigenvector of the shape operator $A$. If $f$ is a function on $M$ satisfying*

$$Xf = (Uf)u(X) \qquad (19.9)$$

*for any $X \in T(M)$, then $f$ is constant on $M$.*

*Proof.* First taking the covariant derivative of (19.9), we obtain

$$Y(Uf)u(X) - X(Uf)u(Y) + (Uf)g((FA + AF)Y, X) = 0.$$

Putting $Y = U$ in the last relation, we get $X(Uf) = U(Uf)u(X)$. Now, the last two equations imply

$$(Uf)g((FA + AF)X, Y) = 0.$$

Finally, since the holomorphic sectional curvature is positive, after using Lemma 19.2, it follows $Uf = 0$ and therefore relation (19.9) implies that $f$ is constant. □

**Lemma 19.4.** *If the shape operator $A$ has exactly two distinct eigenvalues, then they are constant.*

*Proof.* According to Lemma 19.1, $U$ is an eigenvector of the shape operator $A$, that is, $AU = \alpha U$. Hence using relation (19.7) and Lemma 19.3, we conclude that $\alpha$ is constant.

Now denoting another eigenvalue of $A$ by $\lambda$ and the corresponding eigenvector by $X$, and using (19.8), we obtain the following relation:

$$AFX = \frac{2k + \alpha\lambda}{2\lambda - \alpha}FX. \tag{19.10}$$

The last relation implies that $FX$ is an eigenvector corresponding to the eigenvalue

$$\mu = \frac{2k + \alpha\lambda}{2\lambda - \alpha} \tag{19.11}$$

if $X$ is an eigenvector corresponding to the eigenvalue $\lambda$. As $A$ has exactly two distinct eigenvalues, it follows $\mu = \alpha$ or $\mu = \lambda$, and hence $\lambda$ is constant, since $\alpha$ and $k$ are constant. □

In the following, we want to prove that, in the case when $A$ has exactly two distinct eigenvalues, the multiplicity of the eigenvalue $\alpha$ corresponding to the eigenvector $U$ is one. Supposing that $X$ is an eigenvector of $A$ such that $AX = \lambda X$ and $g(X, U) = 0$, it follows, using relation (19.10), that

$$AFX = \mu FX, \quad \text{where} \quad \lambda = \mu, \quad \text{or} \quad \lambda = \alpha, \quad \text{or} \quad \mu = \alpha,$$

since $A$ has exactly two distinct eigenvalues. In the case $\lambda = \alpha$ or $\mu = \alpha$, using relation (19.11), we conclude that the shape operator $A$ has two distinct eigenvalues: $\alpha$ and $\frac{2k+\alpha^2}{\alpha}$. Therefore, the proof is separated into two cases.

First, we suppose that the shape operator $A$ has two distinct eigenvalues: $\alpha$ and $\lambda = \mu$. Then, it follows, using relation (19.11), that

$$\lambda^2 - \alpha\lambda = k. \tag{19.12}$$

Moreover, using relation (19.10), we conclude that $T_\lambda$ is invariant under the action of $F$. Further, suppose that the multiplicity of $\alpha$ is greater than one and let $X \in T_\alpha$ and $g(X, U) = 0$. Then it follows from relation (19.10) that

$$AFX = \frac{2k + \alpha^2}{\alpha} FX = \lambda FX,$$

since $k \neq 0$. Therefore $FX \in T_\lambda$ and consequently, $AF^2X = \lambda F^2X$, that is, $AX = \lambda X$, which is a contradiction since $\lambda \neq \alpha$.

Now we turn to the case when the shape operator $A$ has two distinct eigenvalues $\alpha$ and $\lambda = \frac{2k + \alpha^2}{\alpha}$. Let

$$D_\alpha = \{X | AX = \alpha X, g(X, U) = 0\}, \quad D_\lambda = \{X | AX = \lambda X\}.$$

Since both eigenvalues are constants, it follows that $D_\alpha$ and $D_\lambda$ are $\nu_1$- and $\nu_2$-dimensional distributions, respectively, such that $FD_\alpha = D_\lambda$ and $FD_\lambda = D_\alpha$. Then we have

**Lemma 19.5.** *Assuming that $X$ belongs to $D_\alpha$ (or $D_\lambda$), it follows that $A_a X$, $a = 2, \ldots, p$, belongs to $D_\alpha$ (or $D_\lambda$).*

*Proof.* Since $\xi$ is parallel with respect to the normal connection, using Lemma 15.2, Ricci equation (5.27) and (9.21), we easily see that $[A, A_a] = 0$, which implies our assertion.    □

**Lemma 19.6.** *If $X$ and $Y$ belong to $D_\alpha$ (or $D_\lambda$), then, $\nabla_Y X$ belongs to $D_\alpha$ (or $D_\lambda$), respectively.*

*Proof.* We are going to prove only the case when $X, Y \in D_\alpha$, having in mind that the proof of the case $X, Y \in D_\lambda$ is analogous.

First, we note that $g(\nabla_Y X, U) = 0$ since $X$ is orthogonal to $U$ and $FY \in D_\lambda$. Now, using this fact, it follows $g(A\nabla_Y X, U) = 0$. Further, to prove that $A\nabla_Y X = \alpha\nabla_Y X$, for $X, Y \in D_\alpha$, assume $Z \in T(M)$ is orthogonal to $U$. Then, using the Codazzi equation (19.1), it follows that

$$\begin{aligned}
g(A\nabla_Y X, Z) &= g(\nabla_Y(AX) - g(\nabla_Y A)X, Z) \\
&= \alpha g(\nabla_Y X, Z) - g(X, (\nabla_Y A)Z) \\
&= \alpha g(\nabla_Y X, Z) - g(X, (\nabla_Z A)Y) \\
&\quad - k\{u(Y)g(X, FZ) - u(Z)g(X, FY) + 2g(FZ, Y)u(X)\} \\
&= \alpha g(\nabla_Y X, Z) - g(X, (\nabla_Z A)Y) \\
&= \alpha g(\nabla_Y X, Z) - g(X, \nabla_Z(AY) - A\nabla_Z Y) \\
&= \alpha g(\nabla_Y X, Z),
\end{aligned}$$

since $\alpha$ is constant and $A$, $\nabla_Y A$ are symmetric operators. This completes the proof.    □

From now on, for any $X \in T(M)$, we denote by $X_\alpha$ its $D_\alpha$-component and, analogously, by $X_\lambda$ its $D_\lambda$-component. Then, assuming that $X \in D_\alpha$, $W \in D_\lambda$, after differentiating the relation $g(X, W) = 0$ and using Lemma 19.6, we obtain

**Lemma 19.7.** *Supposing that $X \in D_\alpha$ and $W \in D_\lambda$, it follows that*

$$\nabla_W X \in D_\alpha \oplus \operatorname{span}\{U\}, \quad \nabla_X W \in D_\lambda \oplus \operatorname{span}\{U\},$$

*that is,*

$$\nabla_W X = (\nabla_W X)_\alpha - \lambda g(X, FW)U,$$
$$\nabla_X W = (\nabla_X W)_\lambda - \alpha g(W, FX)U.$$

Now, we are ready to prove

**Lemma 19.8.** *Under the same assumptions as above, with $n > 2p - 1$, $p \geq 2$, if the shape operator $A$ has exactly two distinct eigenvalues, it follows that the multiplicity of the eigenvalue $\alpha$ corresponding to the eigenvector $U$ of the shape operator $A$ is one.*

*Proof.* Assume that the multiplicity of $\alpha$ is greater than one. Then, if $X \in D_\alpha$ and $V, W \in D_\lambda$, taking account of Lemmas 19.4, 19.6, 19.7 and relations (15.25), (15.27), (19.12), we get

$$R(X, V)W = (\nabla_X \nabla_V W)_\lambda - \nabla_V (\nabla_X W)_\lambda - \nabla_{(\nabla_X V)_\lambda} W + (\nabla_{(\nabla_V X)_\alpha} W)_\lambda$$
$$+ \alpha \lambda g(W, FX)FV - (\alpha + \lambda)g(FV, X)\nabla_U W. \qquad (19.13)$$

On the other hand, using Gauss equation (5.22) and (9.21), it follows from Lemma 19.5,

$$R(X, V)W = k\{g(V, W)X - g(FX, W)FV - 2g(FX, V)FW\}$$
$$+ \alpha \lambda g(V, W)X + \sum_{a=2}^{p} g(A_a V, W)A_a X. \qquad (19.14)$$

Further, comparing the $D_\alpha$-components in relations (19.13) and (19.14), we obtain

$$\alpha \lambda g(W, FX)FV - (\alpha + \lambda)g(FV, X)(\nabla_U W)_\alpha = k\{g(V, W)X$$
$$- g(FX, W)FV - 2g(FX, V)FW\} + \alpha \lambda g(V, W)X$$
$$+ \sum_{a=2}^{p} g(A_a W, V)A_a X. \qquad (19.15)$$

Since $\dim D_\lambda = \frac{n-1}{2} > p - 1$, for a fixed $W \in D_\lambda$ and for $a = 2, \ldots, p$, there exists $V \in D_\lambda$ such that $g(A_\alpha W, V) = 0$. Moreover, we can choose $X \in D_\alpha$ in such a way that $g(X, FV) = 0$, since $\dim D_\alpha \geq 2$. Taking these $X$ and $V$ and using relation (19.15), it follows that $\alpha \lambda + k = 0$, which is a contradiction with the relation (19.12), since $k > 0$. $\qquad \square$

*Remark* 19.1. When $M$ is a real hypersurface of a complex projective space $\mathbf{P}^{\frac{n+p}{2}}(\mathbf{C})$ with exactly two distinct eigenvalues of the shape operator $A$, the multiplicity of the eigenvalue $\alpha$ corresponding to the eigenvector $U$ of the shape operator is one.

Namely, in the case of a real hypersurface, relation (19.14) becomes

$$R(X,V)W = k\{g(V,W)X - g(FX,W)FV - 2g(FX,V)FW\} + \alpha\,\lambda\,g(V,W)X,$$

and, consequently, relation (19.15) becomes

$$\alpha\lambda g(W,FX)FV - (\alpha+\lambda)g(FV,X)(\nabla_U W)_\alpha = k\,\{g(V,W)X - g(FX,W)FV$$
$$- 2g(FX,V)FW\} + \alpha\lambda g(V,W)X. \tag{19.16}$$

Moreover, since $\dim D_\alpha \geq 2$, we can choose $X \in D_\alpha$ in such a way that $g(X,FV) = 0$. Therefore, using relation (19.16), we can now proceed analogously as in the proof of Lemma 19.8.

Further, if $M$ is an $n$-dimensional ($n > 2p - 1$, $p \geq 2$) CR submanifold of CR dimension $\frac{n-1}{2}$ of a complex projective space $\mathbf{P}^{\frac{n+p}{2}}(\mathbf{C})$ with two distinct eigenvalues of the shape operator $A$, using the above consideration, we conclude that $A$ can be diagonalized as follows:

$$\begin{pmatrix} \alpha & 0 & \cdots\cdots \\ 0 & \lambda & \cdots\cdots \\ \cdots & \cdots & \cdots\cdots \\ \cdots & \cdots\cdots & \lambda \end{pmatrix}$$

and we may write
$$AX = \lambda X + (\alpha - \lambda)u(X)U. \tag{19.17}$$

Further, let $\mathbf{S}^{n+p+1}$ be the unit sphere in $\mathbf{E}^{n+p+2} = \mathbf{C}^{\frac{n+p+2}{2}}$ and consider the Hopf fibration $\pi : \mathbf{S}^{n+p+1} \to \mathbf{P}^{\frac{n+p}{2}}(\mathbf{C})$. We note that the position vector $z \in \mathbf{S}^{n+p+1}$ in $\mathbf{C}^{\frac{n+p+2}{2}}$ is a unit normal vector to $\mathbf{S}^{n+p+1}$ at $z$ and that $\sqrt{-1}z = -V_z'$, where $V_z'$ is the unit vertical vector at $z$ of the principal fiber bundle $\mathbf{S}^{n+p+1}(\mathbf{P}^{\frac{n+p}{2}}(\mathbf{C}), S^1)$. The fundamental equations for the submersion are given by (9.5) and (9.10).

Let $\Gamma(x,\xi,r)$, $-\infty < r < \infty$, be the geodesic in the complex projective space $\mathbf{P}^{\frac{n+p}{2}}(\mathbf{C})$ parameterized by arc-length $r$ such that

$$\Gamma(x,\xi,0) = x \in \mathbf{P}^{\frac{n+p}{2}}(\mathbf{C}) \quad \text{and} \quad \overrightarrow{\Gamma}(x,\xi,0) = \xi.$$

In terms of the vector representation of $\mathbf{P}^{\frac{n+p}{2}}(\mathbf{C})$, $\Gamma(x,\xi,r)$ can be described as follows. If $w \in \mathbf{S}^{n+p+1}$, such that $\pi(w) = x$ and $\xi^* \in T_w(\mathbf{S}^{n+p+1})$ is the horizontal lift of $\xi$. Then

$$\Gamma(x,\xi,r) = \pi(\cos rw + \sin r\xi^*).$$

We define a map $\Phi_r$ by

$$\Phi_r(x) = \Gamma(x,\xi,r) = \pi(\cos r\, w + \sin r\,\xi^*)$$

and we compute $(\Phi_r)_* X$ for $X \in T_x(M)$. To that purpose, let $\gamma(t)$ be a curve in $M$ with the initial tangent vector $\overrightarrow{\gamma}(0) = X$ and $\gamma^*(t)$ be the horizontal lift of $\gamma(t)$ to $\mathbf{S}^{n+p+1}$ with $\gamma^*(0) = w$ and $\overrightarrow{\gamma}^*(0) = X_w^*$. Then

$$\Phi_r(\gamma(t)) = \pi(\cos r\gamma^*(t)) + \sin r\xi^*(\gamma^*(t)),$$

and therefore, $(\Phi_r)_* X = (\pi_*)_z(\overrightarrow{\eta}(0))$, where $\eta(t)$ is a curve on $\mathbf{S}^{n+p+1}$ defined by

$$\eta(t) = \cos r\, \gamma^*(t) + \sin r\, \xi^* \left(\gamma^*(t)\right)$$

and $z = \cos r\, w + \sin r\, \xi^*$. Considering $\eta(t)$ as a curve in $\mathbf{C}^{\frac{n+p+2}{2}}$, we obtain

$$\overrightarrow{\eta}(0) = \cos r\, X_w^* + \sin r\, \nabla^E_{X_w^*} \xi^*,$$

where $\nabla^E$ is the Euclidean covariant derivative in $\mathbf{C}^{\frac{n+p+2}{2}}$. Since

$$\nabla^E_{X_w^*} \xi^* = \nabla'_{X_w^*} \xi^* - g'(X_w^*, \xi^*)\, w = \nabla'_{X_w^*} \xi^*,$$

using relation (9.10) and the assumption that $\xi$ is parallel with respect to the normal connection, it follows

$$(\pi_*)_w(\nabla'_{X_w^*}\xi^*) = \overline{\nabla}_X \xi = -AX + D_X\xi = -AX.$$

Moreover,

$$\nabla'_{X_w^*}\xi^* = -(AX)^* + g'(\nabla'_{X_w^*}\xi^*, V_w')V_w', \qquad (19.18)$$

where $(AX)^*$ is the horizontal lift of $AX$. On the other hand, using (15.5) and (9.4), we obtain

$$g'(\nabla'_{X_w^*}\xi^*, V_w') = -g'(\xi^*, \nabla'_{X_w^*}V') = -g'(\xi^*, (JX)^*) = -g(U, X). \qquad (19.19)$$

Hence, from (19.18) and (19.19), we get

$$\nabla'_{X_w^*}\xi^* = -(AX)^* - g(U, X)V_w'.$$

Thus, we have

$$\begin{aligned}
(\Phi_r)_* X &= (\pi_*)_z(\cos rX_w^* + \sin r\nabla^E_{X_w^*}\xi^*) \\
&= (\pi_*)_z(\cos rX_w^* + \sin r\nabla'_{X_w^*}\xi^*) \\
&= (\pi_*)_z(\cos rX_w^* - \sin r((AX)^* + g(U, X)V_w')). \qquad (19.20)
\end{aligned}$$

Now, if we put

$$W_z'(X) = \cos r X_w^* - \sin r((AX)^* + g(U, X)V_w'),$$

it follows

$$W_z'(U) = (\cos r - \alpha \sin r)U_w^* + \sin r\, V_w',$$

since $U$ is an eigenvector of the shape operator $A$.

We need to find the horizontal component of $W_z'(U)$, since $r$ is the arc length of the geodesic $\Gamma(x, \xi, r)$ in $\mathbf{P}^{\frac{n+p}{2}}(\mathbf{C})$. Using the fact that

$$V_z' = -\sqrt{-1}z = \cos r V_w' + \sin r U_w^*,$$

we compute

$$g'(W_z'(U), V_z')V_z' = \sin r(2\cos r - \alpha \sin r)(\cos r V_w' + \sin r U_w^*),$$

and therefore, the horizontal part $(W_z'(U))^H$ of $W_z'(U)$ is given by

$$(W_z'(U))^H = (\cos 2r - \frac{\alpha}{2}\sin 2r)(-\sin r\, V_w' + \cos r\, U_w^*). \tag{19.21}$$

From relations (19.20) and (19.21), it follows

$$(\Phi_r)_* U = (\pi_*)(\cos 2r - \frac{\alpha}{2}\sin 2r)(-\sin r\, V_w' + \cos r\, U_w^*), \tag{19.22}$$

$$(\Phi_r)_* X = (\pi_*)(\cos r X_w^* - \sin r)((AX)_w^* + g(U, X)V_w'). \tag{19.23}$$

In particular, if $\alpha = 2\cot 2r$, it follows from (19.22) that $(\Phi_r)_* U = 0$. Further, if $X$ is an eigenvector orthogonal to $U$ and using relation (19.23), we get

$$(\Phi_r)_* X = (\pi_*)_z((\cos r - \lambda \sin r)X_w^*).$$

Since the multiplicity of $\alpha$ is one and $k = 1$, we have $\lambda^2 - \lambda\alpha - 1 = 0$ and therefore,

$$\lambda = \cot 2r \pm \csc 2r.$$

- If $\lambda = \cot 2r + \csc 2r$, then $(\Phi_r)_* X = 0$, for any $x \in M$ and $X \in T_x(M)$, which means that $\Phi_r(M)$ is a single point.

- If $\lambda = \cot 2r - \csc 2r$, then we first note that $\cot 2(r - \frac{\pi}{2}) = \cot 2r$. Hence $(\Phi_{r-\frac{\pi}{2}})_* U = 0$. Now, using $r - \frac{\pi}{2}$ instead of $r$, we obtain $\lambda = \cot r$, and therefore $(\Phi_{r-\frac{\pi}{2}})_* X = 0$. Consequently, $\Phi_{r-\frac{\pi}{2}}(M)$ is a single point.

Finally, using the definition of $\Phi_r$ and $\Phi_{r-\frac{\pi}{2}}$, we can state the following

**Theorem 19.2.** [15] *Let $M$ be an $n$-dimensional ($n > 2p - 1$, $p \geq 2$) CR submanifold of CR dimension $\frac{n-1}{2}$ of a complex projective space $\mathbf{P}^{\frac{n+p}{2}}(\mathbf{C})$. If the shape operator $A$ with respect to the distinguished normal vector field $\xi$ has exactly two distinct eigenvalues, and if $\xi$ is parallel with respect to the normal connection, then there exists a geodesic hypersphere $S$ of $\mathbf{P}^{\frac{n+p}{2}}(\mathbf{C})$ such that $M$ lies on $S$.*

If a real hypersurface $M$ has only two distinct principal curvatures, from Theorem 19.2, there exists such a geodesic hypersphere $S$ such that $M \subset S$. In this case, $\dim M = \dim S = n$ implies that $M$ is an open submanifold of $S$ and if $M$ is complete, $M = S$. Thus we have

**Theorem 19.3.** *[8] If a real hypersurface $M$ of a complex projective space $\mathbf{P}^{\frac{n+1}{2}}(\mathbf{C})$ has two distinct principal curvatures, then $M$ is an open part of geodesic hypersphere. As a consequence of this, $M_{0,q}^{C}$ for $q = \frac{n-1}{2}$ is a geodesic hypersphere.*

# CR submanifolds of maximal CR dimension satisfying the condition $h(FX, Y) + h(X, FY) = 0$

In Sections 20 and 21 we show how some algebraic relations between the naturally induced almost contact structure tensor and the second fundamental form imply the complete classification of CR submanifolds of maximal CR dimension of constant, nonnegative holomorphic sectional curvature.

In this section we study CR submanifolds $M^n$ of maximal CR dimension of $\frac{n+p}{2}$-dimensional complex space forms $\overline{M}$ of nonnegative holomorphic sectional curvature which satisfy the condition

$$h(FX, Y) + h(X, FY) = 0, \tag{20.1}$$

for all $X, Y \in T(M)$. Using Lemma 15.1 and relation (15.14), we obtain

$$\begin{aligned}
h(FX, Y) + h(X, FY) = {} & \{g(AFX, Y) + g(AX, FY)\}\xi \\
& + \sum_{a=1}^{q} \{[g(A_a FX, Y) + g(A_a X, FY)]\xi_a \\
& + [g(A_{a^*} FX, Y) + g(A_{a^*} X, FY)]\xi_{a^*}\}.
\end{aligned}$$

Therefore, since $F$ is skew-symmetric, it follows that the relation (20.1) is equivalent to

$$AF = FA, \tag{20.2}$$
$$A_a F = FA_a, \tag{20.3}$$
$$A_{a^*} F = FA_{a^*}. \tag{20.4}$$

Further, if relation (20.2) holds at a point of the submanifold $M$, using (15.6) and (15.7), we get

$$AU = \alpha U, \tag{20.5}$$

where we have put $\alpha = u(AU)$. Thus, the following lemma holds:

**Lemma 20.1.** *Let $M$ be an $n$-dimensional CR submanifold of maximal CR dimension of a Kähler manifold $\overline{M}$. If the condition (20.2) is satisfied, then*

*U is an eigenvector of the shape operator A with respect to the distinguished normal vector field $\xi$, at any point of M.*

Using (15.15), (20.3) and (15.7), it follows

$$A_{a^*}U = -s_a(U)U. \tag{20.6}$$

From the last relation and (15.15), we obtain

$$s_a(X) = s_a(U)u(X). \tag{20.7}$$

Now, since the condition (20.3) is satisfied, using relations (15.21) and (20.7), it follows $FA_a = 0$. The proof of $FA_{a^*} = 0$ is analogous: using (15.15), (20.3), (15.6) and (15.7), we obtain

$$A_aU = s_{a^*}(U)U \tag{20.8}$$

and using (15.15), it follows

$$s_{a^*}(X) = s_{a^*}(U)u(X). \tag{20.9}$$

Now, since the condition (20.4) is satisfied, using relations (15.22) and (20.9), it follows $FA_{a^*} = 0$. Therefore, using (20.3) and (20.4), we have

$$A_aF = 0 = FA_a, \quad A_{a^*}F = 0 = FA_{a^*}. \tag{20.10}$$

Further, using relations (15.14), (20.9) and (20.10), we obtain

$$A_aX = s_{a^*}(U)u(X)U, \qquad A_{a^*}X = -s_a(U)u(X)U. \tag{20.11}$$

From now on, we suppose that the ambient manifold $\overline{M}$ is a complex space form, that is, a complete Kähler manifold of constant holomorphic sectional curvature $4k$. First, we prove the following

**Lemma 20.2.** *Let M be an n-dimensional CR submanifold of maximal CR dimension of a complex space form $\overline{M}$. If the condition (20.1) is satisfied, then*

*(1) the distinguished normal vector field $\xi$ is parallel with respect to the normal connection, or*

*(2) the ambient manifold $\overline{M}$ is a complex Euclidean space and M is a locally Euclidean space.*

*Proof.* First, using (15.30), (20.7) and (20.10), it follows

$$F\left((\nabla_X A_a)Y - (\nabla_Y A_a)X\right) = s_a(U)(u(Y)FAX - u(X)FAY). \tag{20.12}$$

Now, differentiating the relation (20.10), we obtain

$$(\nabla_X F)A_a Y + F(\nabla_X A_a)Y = 0. \tag{20.13}$$

Further, using (15.20), (20.11) and (20.5), we get

$$(\nabla_X F)A_a Y = s_{a*}(U)u(Y)(AX - \alpha u(X)U), \tag{20.14}$$

and therefore, it follows

$$F((\nabla_X A_a)Y - (\nabla_Y A_a)X) = s_{a*}(U)(u(X)AY - u(Y)AX). \tag{20.15}$$

Using relations (20.12) and (20.15), after replacing $Y$ by $U$, it follows

$$s_a(U)FAX = s_{a*}(U)(\alpha u(X)U - AX), \tag{20.16}$$

since $AU = \alpha U$ (Lemma 20.1).

Now, let $X \in T(M)$, $X \perp U$ be an eigenvector of the shape operator $A$ with respect to distinguished normal vector field $\xi$, namely, $AX = \lambda X$. Then, using (20.2) and (20.16), we obtain

$$\lambda s_{a*}(U)X + \lambda s_a(U)FX = 0. \tag{20.17}$$

Therefore, since $X$ and $FX$ are linearly independent, it follows $D_X \xi = 0$ (since $s_{a*}(U) = 0 = s_a(U)$ and using relations (20.7) and (20.9)), or $\lambda = 0$.

Further, we consider the case $\lambda = 0$. Then the eigenvalues of the shape operator $A$ are $\alpha$ and $0$ and the multiplicity of the eigenvalue $\alpha$ is one. Therefore we may write

$$AY = \alpha u(Y)U \tag{20.18}$$

and it follows

$$FAY = 0. \tag{20.19}$$

Differentiating the last relation and using (15.20), it follows

$$F((\nabla_X A)Y - (\nabla_Y A)X) + u(AY)AX - u(AX)AY = 0. \tag{20.20}$$

Further, using (15.6), (15.7), (15.29) and (20.10), we obtain

$$F((\nabla_X A)Y - (\nabla_Y A)X) = k(u(Y)X - u(X)Y). \tag{20.21}$$

Now, from (20.20) and (20.21), it follows

$$u(AX)AY - u(AY)AX = k(u(Y)X - u(X)Y). \tag{20.22}$$

Replacing $Y$ by $FY$ in the last relation, we obtain $k = 0$, that is, the ambient manifold is a complex Euclidean space.

Finally, using (20.11), (20.18) and the Gauss equation (15.28), it follows $R(X, Y)Z = 0$, that is, $M$ is a locally Euclidean space.    □

Now we consider the case when the ambient manifold $\overline{M}^{\frac{n+p}{2}}$ is a complex space form and when the CR submanifold $M$ of maximal CR dimension satisfies (1) in Lemma 20.2, that is, the case when the distinguished normal vector $\xi$ is parallel with respect to the normal connection $D$ which is induced from $\overline{\nabla}$, namely,

$$D_X\xi = \sum_{a=1}^{q}\{s_a(X)\xi_a + s_{a^*}(X)\xi_{a^*}\} = 0, \tag{20.23}$$

from which it follows

$$s_a = s_{a^*} = 0, \quad a = 1,\dots,q. \tag{20.24}$$

Therefore, using relations (15.14) and (15.15), we obtain

$$A_{a^*} = FA_a, \quad A_a = -FA_{a^*}, \quad a = 1,\dots,q, \tag{20.25}$$

$$A_{a^*}U = 0 = A_aU, \quad a = 1,\dots,q, \quad q = \frac{p-1}{2}. \tag{20.26}$$

Moreover, using (15.21) and (15.22), we obtain

$$A_aF + FA_a = 0, \quad A_{a^*}F + FA_{a^*} = 0, \quad a = 1,\dots,q. \tag{20.27}$$

Further, we continue our investigation of the condition (20.1). Since this condition is equivalent to (20.2), (20.3) and (20.4), using (20.25) and (20.27), it follows

$$A_a = 0 = A_{a^*}, \quad a = 1,\dots,q.$$

Namely, we proved that the following lemma holds:

**Lemma 20.3.** *Let $M$ be a complete $n$-dimensional CR submanifold of CR dimension $\frac{n-1}{2}$ of a complex space form. If the distinguished normal vector field $\xi$ is parallel with respect to the normal connection and if the condition (20.1) is satisfied, then $A_a = 0 = A_{a^*}$, $a = 1,\dots,q$, where $A_a$, $A_{a^*}$ are the shape operators for the normals $\xi_a$, $\xi_{a^*}$, respectively.*

Combining this result with Theorem 14.3, we prove

**Theorem 20.1.** *Let $M$ be a complete $n$-dimensional CR submanifold of CR dimension $\frac{n-1}{2}$ of a complex space form of nonnegative holomorphic sectional curvature. If the distinguished normal vector field $\xi$ is parallel with respect to the normal connection and if the condition (20.1) is satisfied, then there exists a totally geodesic complex space form $M'$ of $\overline{M}$ such that $M$ is a real hypersurface of $M'$.*

*Proof.* First, we put $N_0(x) = \{\xi \in T_x^{\perp}(M)|A_{\xi} = 0\}$ and let $H_0(x)$ be the maximal $J$-invariant subspace of $N_0(x)$, that is, $H_0(x) = JN_0(x) \cap N_0(x)$. Then, using Lemma 20.3 and Theorem 14.3, the submanifold $M$ may be regarded as a real hypersurface of $\mathbf{C}^{\frac{n+1}{2}}$, $\mathbf{P}^{\frac{n+1}{2}}(\mathbf{C})$, which are totally geodesic submanifolds in $\mathbf{C}^{\frac{n+p}{2}}$, $\mathbf{P}^{\frac{n+p}{2}}(\mathbf{C})$, because here we consider only the case when the ambient manifolds have nonnegative holomorphic sectional curvature. $\square$

In what follows we denote $\mathbf{C}^{\frac{n+1}{2}}$, $\mathbf{P}^{\frac{n+1}{2}}(\mathbf{C})$, by $M'$ and by $\imath_1$ the immersion of $M$ into $M'$ and by $\imath_2$ the totally geodesic immersion of $M'$ into $\mathbf{C}^{\frac{n+p}{2}}$, $\mathbf{P}^{\frac{n+p}{2}}(\mathbf{C})$, respectively. Then, from the Gauss formula (5.1), it follows that

$$\nabla'_X \imath_1 Y = \imath_1 \nabla_X Y + g(A'X,Y)\xi',$$

where $A'$ is the corresponding shape operator and $\xi'$ is a unit normal vector field to $M$ in $M'$. Consequently, using the Gauss formula (5.1) and $\imath = \imath_2 \cdot \imath_1$, we derive

$$\overline{\nabla}_X \imath_2 \circ \imath_1 Y = \imath_2 \nabla'_X \imath_1 Y + \bar{h}(\imath_1 X, \imath_1 Y)$$
$$= \imath_2(\imath_1 \nabla_X Y + g(A'X,Y)\xi'), \qquad (20.28)$$

since $M'$ is totally geodesic in $\mathbf{C}^{\frac{n+p}{2}}$, $\mathbf{P}^{\frac{n+p}{2}}(\mathbf{C})$. Further, comparing relation (20.28) with relation (5.1), it follows that $\xi = \imath_2 \xi'$ and $A = A'$. As $M'$ is a complex submanifold of $\mathbf{C}^{\frac{n+p}{2}}$, $\mathbf{P}^{\frac{n+p}{2}}(\mathbf{C})$ with the induced complex structure $J'$, we have $J \imath_2 X' = \imath_2 J' X'$, $X' \in T(M')$. Thus, from (15.2) it follows that

$$J \imath X = \imath_2 J' \imath_1 X = \imath F' X + \nu'(X) \imath_2 \xi' = \imath F' X + \nu'(X)\xi \qquad (20.29)$$

and therefore, we conclude that $F = F'$ and $\nu' = u$ and since $M$, for which condition (20.2) is fulfilled, is a real hypersurface of $\mathbf{P}^{\frac{n+1}{2}}(\mathbf{C})$, $\mathbf{C}^{\frac{n+1}{2}}$, we may use Theorems 16.3 and 16.4, and therefore, using Lemma 20.2 prove the following theorem:

**Theorem 20.2.** [23] *Let $M$ be a complete $n$-dimensional CR submanifold of maximal CR dimension of an $\frac{n+p}{2}$-dimensional complex space form $\overline{M}$ of nonnegative holomorphic sectional curvature. If the condition*

$$h(FX,Y) + h(X,FY) = 0, \quad \text{for all} \quad X,Y \in T(M)$$

*is satisfied, where $F$ is the induced almost contact structure and $h$ is the second fundamental form of $M$, then one of the following statements holds:*

(1) *$M$ is a complete $n$-dimensional CR submanifold of CR dimension $\frac{n-1}{2}$ of a complex Euclidean space $\mathbf{C}^{\frac{n+p}{2}}$, and then $M$ is isometric to $\mathbf{E}^n$, $\mathbf{S}^n$, $\mathbf{S}^{2k+1} \times \mathbf{E}^{n-2k-1}$;*

(2) *$M$ is a complete $n$-dimensional CR submanifold of CR dimension $\frac{n-1}{2}$ of a complex projective space $\mathbf{P}^{\frac{n+p}{2}}(\mathbf{C})$, and then $M$ is isometric to $M_{k,l}^C$, for some $k$, $l$ satisfying $2k + 2l = n - 1$.*

*Remark 20.1.* When $\overline{M}$ is a complex hyperbolic space $\mathbf{H}^{\frac{n+p}{2}}(\mathbf{C})$, the complete classification is given in [23].

# Contact CR submanifolds of maximal CR dimension

In this section we study CR submanifolds $M^n$ of maximal CR dimension of a complex space form $\overline{M}^{\frac{n+p}{2}}$ which satisfy the condition

$$h(FX, Y) - h(X, FY) = g(FX, Y)\eta, \quad \eta \in T^{\perp}(M) \tag{21.1}$$

for all $X, Y \in T(M)$, where $\eta$ does not have zero points.

For now, let $(\overline{M}, J, \overline{g})$ be a Kähler manifold. According to Lemma 15.1 and setting

$$\eta = \rho\xi + \sum_{a=1}^{q}(\rho^a \xi_a + \rho^{a^*} \xi_{a^*}),$$

we conclude that the condition (21.1) is equivalent to

$$AFX + FAX = \rho FX, \tag{21.2}$$

$$A_a FX + FA_a X = \rho^a FX, \tag{21.3}$$

$$A_{a^*} FX + FA_{a^*} X = \rho^{a^*} FX, \tag{21.4}$$

for all $a = 1, \ldots, q$, $q = \frac{p-1}{2}$, since $F$ is a skew-symmetric endomorphism acting on $T(M)$. Here we also used relation (15.14).

Combining relations (4.4), (15.2), (15.26) and (21.2), we compute

$$\begin{aligned} du(X, Y) &= (\nabla_X u)(Y) - (\nabla_Y u)(X) \\ &= g((FA + AF)X, Y) \\ &= \rho\, g(FX, Y). \end{aligned} \tag{21.5}$$

Then, since $F$ has rank $n - 1$, we conclude

$$u \wedge du \wedge \cdots \wedge du \neq 0. \tag{21.6}$$

We recall the definition of a contact manifold. A manifold $M^{2m+1}$ is said to be a *contact manifold* if it carries a global one-form $u$ such that

M. Djorić, M. Okumura, *CR Submanifolds of Complex Projective Space*, Developments in Mathematics 19, DOI 10.1007/978-1-4419-0434-8_21, © Springer Science+Business Media, LLC 2010

$$u \wedge (du)^m \neq 0 \qquad (21.7)$$

everywhere on $M$. The one-form $u$ is called the *contact form*.

Relation (21.6) now proves

**Proposition 21.1.** *If $M$ is a CR submanifold of maximal CR dimension of a Kähler manifold, which satisfies the condition (21.2), for $\rho \neq 0$, then $M$ is a contact manifold.*

Further, using (21.5), (4.2) and (15.2), we obtain

$$\begin{aligned} du(X,Y) &= \rho\, g(FX,Y) = \rho \bar{g}(\imath F X, \imath Y) \\ &= \rho \bar{g}(J\imath X, \imath Y) = \rho \Omega(\imath_* X, \imath_* Y) \\ &= \rho \imath^* \Omega(X,Y), \end{aligned} \qquad (21.8)$$

where $\imath^*$ is the pull-back map which commutes with the exterior derivative. Consequently, using (21.8), we compute $d\rho = 0$ since

$$0 = d^2 u = d\rho \wedge \imath^* \Omega + \rho d\imath^* \Omega = d\rho \wedge \imath^* \Omega + \rho \imath^* d\Omega = d\rho \wedge \imath^* \Omega,$$

which yields

**Lemma 21.1.** *If $M^n$, $n > 3$ is a CR submanifold of maximal CR dimension of a Kähler manifold, which satisfies the condition (21.2), then $\rho$ is constant.*

**Lemma 21.2.** *Let $M$ be an $n$-dimensional CR submanifold of maximal CR dimension of a Kähler manifold $\overline{M}$. If the condition (21.1) is satisfied, it follows*

$$FA_a + A_a F = 0, \quad FA_{a^*} + A_{a^*} F = 0, \qquad (21.9)$$

*that is, $\rho^a = 0$, $\rho^{a^*} = 0$, $a = 1, \ldots, q$.*

*Proof.* Since the condition (21.1) is equivalent to (21.2), (21.3) and (21.4), using (15.21) and (15.22), we get

$$\rho^a g(FX,Y) = s_a(X)u(Y) - s_a(Y)u(X), \qquad (21.10)$$
$$\rho^{a^*} g(FX,Y) = s_{a^*}(X)u(Y) - s_{a^*}(Y)u(X). \qquad (21.11)$$

Next, we put $Y = U$ in (21.10) and (21.11) and use (15.7) to obtain

$$s_a(X) = s_a(U)u(X), \quad s_{a^*}(X) = s_{a^*}(U)u(X). \qquad (21.12)$$

Substituting (21.12) into (15.21) and (15.22), we get (21.9). Finally, using (21.3) and (21.4), we have $\rho^a = 0$ and $\rho^{a^*} = 0$, for $a = 1, \ldots, q$. $\qquad \square$

*Remark* 21.1. From Lemma 21.2, together with (21.1), we conclude that $\rho$ does not have zero points.

Further, using (21.9) and (15.7), it follows

$$FA_aU = 0, \quad FA_{a^*}U = 0. \tag{21.13}$$

Therefore, using (21.13) and (15.6), we obtain

$$A_aU = s_{a^*}(U)U, \quad A_{a^*}U = -s_a(U)U. \tag{21.14}$$

Further, if relation (21.2) holds at a point of the submanifold $M$, using (15.6) and (15.7), we get that $U$ is an eigenvector of the shape operator $A$ with respect to distinguished normal vector field $\xi$, at any point of $M$, namely,

$$AU = \alpha U, \tag{21.15}$$

where we have put $\alpha = u(AU)$. Namely, we proved

**Lemma 21.3.** *Let $M$ be an $n$-dimensional CR submanifold of maximal CR dimension of a Kähler manifold $\overline{M}$. If the condition (21.1) is satisfied, then $U$ is an eigenvector of the shape operator $A$ with respect to distinguished normal vector field $\xi$, at any point of $M$.*

From now on we assume that the ambient manifold $\overline{M}$ is a complex space form.

Using Ricci-Kühne formula (5.24), Gauss equation (5.22), relations (15.2) and (15.5), we obtain

$$\begin{aligned}
0 = \bar{g}(\overline{R}(\imath X, \imath Y)\xi_a, \xi) = {} & g(AA_aX, Y) - g(A_aAX, Y) \\
& + (\nabla_X s_a)(Y) - (\nabla_Y s_a)(X) \\
& + \sum_{b=1}^{q}[s_b(Y)s_{ba}(X) + s_{b^*}(Y)s_{b^*a}(X) \\
& - s_b(X)s_{ba}(Y) - s_{b^*}(X)s_{b^*a}(Y)]. \tag{21.16}
\end{aligned}$$

We now prove the following extremely useful

**Lemma 21.4.** *Let $M$ be a complete $n$-dimensional CR submanifold of CR dimension $\frac{n-1}{2}$ of a complex space form. If the condition (21.1) is satisfied, then the distinguished normal vector field $\xi$ is parallel with respect to the normal connection.*

*Proof.* Let us compute $g((\nabla_X A_{a^*})Y - (\nabla_Y A_{a^*})X, U)$ in the following two ways. First, differentiating the relation (15.15) and using (15.25), (15.27), (21.14) and (21.15), we obtain

$$\begin{aligned}
g((\nabla_X A_{a^*})Y, U) &= g((\nabla_X F)A_aY, U) + g(F(\nabla_X A_a)Y, U) - (\nabla_X s_a)(Y) \\
&= -g(A_aAX, Y) + \alpha s_{a^*}(U)u(X)u(Y) - (\nabla_X s_a)(Y). \tag{21.17}
\end{aligned}$$

Reversing X and Y and subtracting thus yields

$$g((\nabla_X A_{a^*})Y - (\nabla_Y A_{a^*})X, U) = g((AA_a - A_a A)X, Y)$$
$$- (\nabla_X s_a)(Y) + (\nabla_Y s_a)(X). \tag{21.18}$$

Substituting (15.31) into (21.18) and using (21.15), we obtain

$$g((AA_a - A_a A)X, Y) - (\nabla_X s_a)(Y) + (\nabla_Y s_a)(X) = \tag{21.19}$$
$$\sum_{b=1}^{q} \{ s_{a^*b}(X) g(A_b Y, U) - s_{a^*b}(Y) g(A_b X, U) \}$$
$$+ \sum_{b=1}^{q} \{ s_{a^*b^*}(X) g(A_{b^*} Y, U) - s_{a^*b^*}(Y) g(A_{b^*} X, U) \}.$$

Now, using (15.15), (15.17), (15.19), relations (21.16) and (21.19) yield

$$g((AA_a - A_a A)X, Y) = 0, \quad \text{for all} \quad X, Y \in T(M). \tag{21.20}$$

Next, differentiating the second relation of (21.14) and using (15.27) and (21.2), we obtain

$$g((\nabla_X A_{a^*})Y - (\nabla_Y A_{a^*})X, U) + g((A_{a^*} FA + AF A_{a^*})X, Y) \tag{21.21}$$
$$= Y(s_a(U))u(X) - X(s_a(U))u(Y) - \rho \, s_a(U) \, g(FX, Y).$$

Since $g(FX, FY) = g(X, Y) - u(X)u(Y)$, using (15.15), (15.17), (21.15) and (21.12), we compute

$$g((A_{a^*} FA + AF A_{a^*})X, Y) = g(AX, A_a Y) - u(AX)u(A_a Y)$$
$$- g(AY, A_a X) + u(AY)u(A_a X)$$
$$= g(AX, A_a Y) - g(AY, A_a X)$$
$$+ g(AY, U)s_{a^*}(U)u(X) - g(AX, U)s_{a^*}(U)u(Y)$$
$$= g((A_a A - AA_a)X, Y). \tag{21.22}$$

Codazzi equation (15.31), together with (15.18), (15.17) and (21.15), yields

$$g((\nabla_X A_{a^*})Y - (\nabla_Y A_{a^*})X, U) = \sum_{b=1}^{q} \{ s_{a^*b}(X)s_{b^*}(Y) - s_{a^*b}(Y)s_{b^*}(X) \}$$
$$+ \sum_{b=1}^{q} \{ s_{a^*b^*}(Y)s_b(X) - s_{a^*b^*}(X)s_b(Y) \}. \tag{21.23}$$

Now, using (21.22) and (21.23), relation (21.21) reads

$$\sum_{b=1}^{q} \{ s_{a^*b}(X)s_{b^*}(Y) - s_{a^*b}(Y)s_{b^*}(X) - s_{a^*b^*}(X)s_b(Y) + s_{a^*b^*}(Y)s_b(X) \}$$
$$+ g((A_a A - AA_a)X, Y) = Y(s_a(U))u(X) - X(s_a(U))u(Y)$$
$$- \rho s_a(U)g(FX, Y). \tag{21.24}$$

Further, replacing $Y$ by $U$ in relation (21.24) and using (21.12), we obtain

$$X(s_a(U)) = U(s_a(U))u(X) - \sum_{b=1}^{q}\{s_{a*b}(X)s_{b*}(U) - s_{a*b*}(X)s_b(U)$$
$$- u(X)[s_{a*b}(U)s_{b*}(U) + s_{a*b*}(U)s_b(U)]\}, \tag{21.25}$$

since, using (21.14) and (21.15), we compute $g((A_aA - AA_a)X,U) = 0$. Combining relation (21.25) with (21.24) and using (21.12), we get

$$g((AA_a - A_aA)X,Y) = \rho s_a(U)g(FX,Y). \tag{21.26}$$

Thus (21.20) and (21.26) imply $s_a(U) = 0$ and consequently, from (21.12) we conclude $s_a(X) = 0$. In entirely the same way, we obtain $s_{a*} = 0$, which completes the proof.                                                              □

*Remark 21.2.* A slight change in the proof of (21.20), implies

$$g((AA_{a*} - A_{a*}A)X,Y) = 0, \quad \text{for all} \quad X,Y \in T(M). \tag{21.27}$$

Further, using (21.15) and (15.27), we calculate directly

$$(\nabla_X A)U = (X\alpha)U + \alpha FAX - AFAX$$

and, since $A$ is a symmetric operator, taking the inner product of $(\nabla_X A)U$ with $Y$, we obtain

$$g((\nabla_X A)Y,U) = (X\alpha)u(Y) + \alpha g(FAX,Y) - g(AFAX,Y). \tag{21.28}$$

Then, interchanging $X$ and $Y$ in (21.28) and subtracting, gives

$$g((\nabla_X A)Y - (\nabla_Y A)X,U) = (X\alpha)u(Y) - (Y\alpha)u(X) \tag{21.29}$$
$$+ \alpha(g(FAX,Y) - g(FAY,X)) - g(AFAX,Y) + g(AFAY,X).$$

Since $M$ is a CR submanifold of a complex space form, the Codazzi equation (15.29), Lemma 21.4 and relation (21.29) imply

$$(X\alpha)u(Y) - (Y\alpha)u(X) + \alpha g((FA + AF)X,Y)$$
$$- 2g(AFAX,Y) = -2kg(FX,Y). \tag{21.30}$$

If we set $Y = U$ in (21.30), we get $X\alpha = \beta u(X)$, where $\beta = U\alpha$, that is,

$$\operatorname{grad}\alpha = \beta U. \tag{21.31}$$

Taking the covariant derivative of (21.31), reversing $X$ and $Y$ and subtracting the two equations and using (21.2), we obtain

$$0 = g(\nabla_Y \operatorname{grad}\alpha, X) - g(\nabla_X \operatorname{grad}\alpha, Y)$$
$$= (Y\beta)u(X) - (X\beta)u(Y) + \rho\beta g(FY,X). \tag{21.32}$$

Replacing $Y$ by $U$ in (21.32), we get $X\beta = (U\beta)u(X)$ and substituting this into (21.32), we have $\rho\beta g(FX, Y) = 0$. Since $\rho \neq 0$, we conclude $\beta = 0$. This, together with (21.31), implies that $\alpha$ is constant and we have thus proved

**Lemma 21.5.** *Let $M$ be a complete $n$-dimensional CR submanifold of CR dimension $\frac{n-1}{2}$ of a complex space form. If the condition (21.1) is satisfied, then the eigenvalue $\alpha = u(AU)$, corresponding to $U$, is constant.*

Consequently, since $\alpha$ is constant, relation (21.30) becomes

$$\alpha(FA + AF)X - 2AFAX = -2kFX. \tag{21.33}$$

Applying $F$ to relation (21.33) and using (21.2), we obtain

$$2A^2X - 2\rho AX + (\alpha\rho + 2k)X - (2\alpha^2 - \alpha\rho + 2k)u(X)U = 0, \tag{21.34}$$

and we are thus led to the following

**Lemma 21.6.** *Let $M$ be a complete $n$-dimensional CR submanifold of CR dimension $\frac{n-1}{2}$ of a complex space form. If the condition (21.1) is satisfied, then $A$ has at most three distinct eigenvalues and they are constant.*

*Proof.* On account of Lemma 21.4 and Lemma 21.5, we know that $\alpha$ is a constant eigenvalue, corresponding to the eigenvector $U$. Since $A$ is a symmetric operator, let $X$ be another eigenvector with the corresponding eigenvalue $\lambda$. Then, according to (21.34), it follows

$$2\lambda^2 - 2\rho\lambda + (\alpha\rho + 2k) = 0 \tag{21.35}$$

since $X \perp U$. Thus $A$ has at most three distinct eigenvalues which are all constant. □

Now, using Lemma 21.4 and relation (15.15), it follows $A_{a*} = FA_a$. Combining this with relation (21.27) and (21.9), we obtain

$$A_aFAY + AFA_aY = 0,$$

for any tangent vector $Y$. Therefore, using (21.2), (21.9) and (21.20), we conclude

$$\rho A_aFY - 2A_aAFY = 0. \tag{21.36}$$

For another eigenvector $X$, orthogonal to $U$, with the corresponding eigenvalue $\lambda$, since $X$ can be written as $X = FY$ and $AFY = \lambda FY = \lambda X$, we can rewrite (21.36) as

$$(\rho - 2\lambda)A_a X = 0. \tag{21.37}$$

First we consider the case when one of the eigenvalues $\lambda$ is different from $\frac{\rho}{2}$. It follows from (21.37) that $A_a X = 0$ for all $X \perp U$. Further, using (15.18) and Lemma 21.4, it follows

$$g(A_a U, Y) = s_{a^*}(Y) = 0, \quad \text{for all } Y \in T(M)$$

and therefore $A_a U = 0$. Hence, taking into account that $A_a X = 0$, for all $X$ orthogonal to $U$, it follows $A_a = 0$ for $a = 1, \ldots, q$.

Using this procedure, the proof of $A_{a^*} = 0$, $a = 1, \ldots, q$ is essentially the same, and so we omit it. Hence, the following lemma holds:

**Lemma 21.7.** *Let $M$ be a complete $n$-dimensional CR submanifold of CR dimension $\frac{n-1}{2}$ of an $\frac{n+p}{2}$-dimensional complex space form $\overline{M}$. If the condition (21.1) is satisfied and $\rho \neq 2\lambda$, where $X$ is another eigenvector of $A$, orthogonal to $U$, with the corresponding eigenvalue $\lambda$, then $A_a = 0 = A_{a^*}$, $a = 1, \ldots, q$, $q = \frac{p-1}{2}$, where $A$, $A_a$, $A_{a^*}$ are the shape operators for the normals $\xi$, $\xi_a$, $\xi_{a^*}$, respectively.*

Making use of this result, we prove

**Theorem 21.1.** *Let $M$ be a complete $n$-dimensional CR submanifold of CR dimension $\frac{n-1}{2}$ of a complex projective space $\mathbf{P}^{\frac{n+p}{2}}(\mathbf{C})$ (respectively a complex Euclidean space $\mathbf{C}^{\frac{n+p}{2}}$). If the condition (21.1) is satisfied and $\rho \neq 2\lambda$, where $X$ is another eigenvector of $A$, orthogonal to $U$, with the corresponding eigenvalue $\lambda$, then there exists a totally geodesic complex projective subspace $\mathbf{P}^{\frac{n+1}{2}}(\mathbf{C})$ (respectively complex subspace $\mathbf{C}^{\frac{n+1}{2}}$) of $\mathbf{P}^{\frac{n+p}{2}}(\mathbf{C})$ (respectively $\mathbf{C}^{\frac{n+p}{2}}$) such that $M$ is real hypersurface of $\mathbf{P}^{\frac{n+1}{2}}(\mathbf{C})$ (respectively $\mathbf{C}^{\frac{n+1}{2}}$).*

*Proof.* First, let us define

$$N_0(x) = \{\xi \in T_x^\perp(M) | A_\xi = 0\}$$

and let $H_0(x)$ be the maximal $J$-invariant subspace of $N_0(x)$, that is,

$$H_0(x) = JN_0(x) \cap N_0(x).$$

Then, using Lemma 21.10, it follows

$$N_0(x) = \text{span}\{\xi_1(x), \ldots, \xi_q(x), \xi_{1^*}(x), \ldots, \xi_{q^*}(x)\}.$$

Since $J\xi_a = \xi_{a^*}$, we have $JN_0(x) = N_0(x)$ and consequently

$$H_0(x) = JN_0(x) \cap N_0(x) = \text{span}\{\xi_1(x), \ldots, \xi_q(x), \xi_{1^*}(x), \ldots, \xi_{q^*}(x)\}.$$

Hence the orthogonal complement $H_1(x)$ of $H_0(x)$ in $T_x^\perp(M)$ is spanned by $\xi$, which is parallel with respect to the normal connection, by Lemma 21.4. Therefore, we can apply the codimension reduction theorems (see Section 14).

If $\overline{M}$ is a complex projective space, applying Theorem 14.3 we conclude that there exists a real $(n+1)$-dimensional totally geodesic complex projective subspace $\mathbf{P}^{\frac{n+1}{2}}(\mathbf{C})$, such that $M$ is a real hypersurface of it.

If $\overline{M}$ is a complex Euclidean space, applying Theorem 14.1 we conclude that there exists a real $(n+1)$-dimensional totally geodesic Euclidean space $\mathbf{E}^{n+1}$, such that $M$ is a real hypersurface of it. Since $T(\mathbf{E}^{n+1}) = T(M) \oplus \xi$, we have

$$X' = \imath X + a\xi, \quad \text{for} \quad X' \in T(\mathbf{E}^{n+1}), \quad X \in T(M),$$

Then, by (15.2) and (15.5), it follows

$$JX' = J\imath X + Ja\xi = \imath FX + u(X)\xi - a\imath U \in T(M) \oplus \xi = T(\mathbf{E}^{n+1}).$$

and we conclude that $\mathbf{E}^{n+1}$ is $J$-invariant and therefore complex. Consequently, there exists a real $(n+1)$-dimensional totally geodesic complex Euclidean subspace of $\mathbf{C}^{\frac{n+1}{2}}$, such that $M$ is its real hypersurface.     □

Therefore, we can apply the results of real hypersurface theory. Namely, using Theorem 21.1, the submanifold $M$ can be regarded as a real hypersurface of $\mathbf{P}^{\frac{n+1}{2}}(\mathbf{C})$ (respectively $\mathbf{C}^{\frac{n+1}{2}}$), which is a totally geodesic submanifold in $\mathbf{P}^{\frac{n+p}{2}}(\mathbf{C})$ (respectively $\mathbf{C}^{\frac{n+p}{2}}$). In what follows we denote by $\imath_1$ the immersion of $M$ into $\mathbf{P}^{\frac{n+1}{2}}(\mathbf{C})$ (respectively $\mathbf{C}^{\frac{n+1}{2}}$), and by $\imath_2$ the totally geodesic immersion of $\mathbf{P}^{\frac{n+1}{2}}(\mathbf{C})$ (respectively $\mathbf{C}^{\frac{n+1}{2}}$) into $\mathbf{P}^{\frac{n+p}{2}}(\mathbf{C})$ (respectively $\mathbf{C}^{\frac{n+p}{2}}$). Then, from the Gauss formula (5.1), it follows

$$\nabla'_X \imath_1 Y = \imath_1 \nabla_X Y + g(A'X, Y)\xi',$$

where $\xi'$ is a unit normal vector field to $M$ in $\mathbf{P}^{\frac{n+1}{2}}(\mathbf{C})$ (respectively $\mathbf{C}^{\frac{n+1}{2}}$) and $A'$ is the corresponding shape operator. Consequently, by using the Gauss formula and $\imath = \imath_2 \cdot \imath_1$, we derive

$$\overline{\nabla}_X \imath_2 \circ \imath_1 Y = \imath_2 \nabla'_X \imath_1 Y + \overline{h}(\imath_1 X, \imath_1 Y) = \imath_2(\imath_1 \nabla_X Y + g(A'X, Y)\xi') \quad (21.38)$$

since $\mathbf{P}^{\frac{n+1}{2}}(\mathbf{C})$(respectively $\mathbf{C}^{\frac{n+1}{2}}$) is totally geodesic in $\mathbf{P}^{\frac{n+p}{2}}(\mathbf{C})$ (respectively $\mathbf{C}^{\frac{n+p}{2}}$). Further, comparing relation (21.38) with relation (5.1), it follows that $\xi = \imath_2\xi'$ and $A = A'$. As $\mathbf{P}^{\frac{n+1}{2}}(\mathbf{C})$ (respectively $\mathbf{C}^{\frac{n+1}{2}}$) is a complex submanifold of $\mathbf{P}^{\frac{n+p}{2}}(\mathbf{C})$ (respectively $\mathbf{C}^{\frac{n+p}{2}}$), with the induced complex structure $J'$, we have

$$J\imath_2 X' = \imath_2 J'X', \quad X' \in T(\mathbf{P}^{\frac{n+1}{2}}(\mathbf{C})).$$

Thus, from (15.2) it follows

$$J\imath X = \imath_2 J'\imath_1 X = \imath F'X + \nu'(X)\imath_2\xi' = \imath F'X + \nu'(X)\xi$$

and therefore, we conclude that $F = F'$ and $\nu' = u$.

Further, we suppose that the ambient manifold is a complex Euclidean space. Then, we have

**Theorem 21.2.** [21] *Let $M$ be a complete $n$-dimensional CR submanifold of maximal CR dimension of a complex Euclidean space $\mathbf{C}^{\frac{n+p}{2}}$. If the condition*

$$h(FX,Y) - h(X,FY) = g(FX,Y)\eta, \quad \eta \in T^{\perp}(M)$$

*is satisfied, where $F$ and $h$ are the induced almost contact structure and the second fundamental form of $M$, respectively, then $M$ is congruent to one of the following:*

$$\mathbf{S}^n, \qquad \mathbf{S}^{\frac{n-1}{2}} \times \mathbf{E}^{\frac{n+1}{2}},$$

*or there exists a geodesic hypersphere $\mathbf{S}^{n+p-1}(\frac{1}{|\alpha|})$ of $\mathbf{C}^{\frac{n+p}{2}}$ such that $M$ is an invariant submanifold by the almost contact structure $F'$ of the hypersphere $\mathbf{S}^{n+p-1}(\frac{1}{|\alpha|})$.*

*Proof.* We first consider the case $\rho \neq 2\lambda$. Then, by Theorem 21.1, $M$ and $A$ are respectively regarded as a hypersurface of $\mathbf{C}^{\frac{n+1}{2}}$ and its shape operator in $\mathbf{C}^{\frac{n+1}{2}}$. Hence, using Lemma 21.6 and Theorem 11.4, we conclude that $M$ must be one of $\mathbf{S}^n$, $\mathbf{E}^n$ and $\mathbf{S}^r \times \mathbf{E}^{n-r}$. However, from (21.2), it follows that $\mathbf{E}^n$ cannot satisfy the condition (21.1). Now, let us determine the dimension $r$ of the component $\mathbf{S}^r$ of $\mathbf{S}^r \times \mathbf{E}^{n-r}$. Using Theorem 11.3, it follows that one principal curvature must be 0, which is a solution of (21.35). Since $k = 0$ and $\rho \neq 0$, $\alpha = 0$ and the principal curvatures are $\rho$ and 0. Moreover, it is easily seen that $FD_0 = D_\rho$, $FD_\rho = D_0$, where

$$D_0 = \{X \in T(M)|\, AX = 0,\, X \perp U\},$$
$$D_\rho = \{X \in T(M)|\, AX = \rho X\}.$$

Hence, $\dim D_0 = \dim D_\rho = \frac{n-1}{2}$. The components $\mathbf{S}^r$ and $\mathbf{E}^{n-r}$ are the integral manifolds of $D_\rho$ and $D_0 \oplus \text{span}\{U\}$, $r = \frac{n-1}{2}$, respectively.

Next, we consider the case when $\rho = 2\lambda$. Then, substituting this into (21.35), we have $\lambda(\lambda - \alpha) = 0$, since $k = 0$. Moreover, using the fact that $\lambda = \frac{\rho}{2} \neq 0$, we conclude that the shape operator $A$ has only one eigenvalue $\alpha$, and therefore, $M$ lies on a hypersphere $\mathbf{S}^{n+p-1}(\frac{1}{|\alpha|})$.

Finally we prove that $M$ is an invariant submanifold by the almost contact structure $F'$ of $\mathbf{S}^{n+p-1}$. Denoting by $\imath_0$ the immersion of $\mathbf{S}^{n+p-1}$ into $\mathbf{C}^{\frac{m+p}{2}}$, we have

$$J\imath_0 \tilde{X} = \imath_0 \tilde{F}\tilde{X} + \tilde{u}(\tilde{X})\xi,$$

for any $\tilde{X} \in T(S)$, since $\xi$ is also unit normal to $\mathbf{S}^{n+p-1}$ in $\mathbf{C}^{\frac{m+p}{2}}$. Here $(\tilde{F}, \tilde{u}, \xi)$ is the induced contact structure of $\mathbf{S}^{n+p-1}$. If $\tilde{X} = \imath'X$, where $\imath'$ is immersion of $M$ into $\mathbf{S}^{n+p-1}$, we have

$$J\imath_0 \circ \imath'X = \imath_0 \tilde{F}\imath'X + \tilde{u}(\imath'X)\xi.$$

Comparing the tangential and normal part, and using (15.2), for $\imath = \imath_0 \cdot \imath'$, we conclude $u(X) = \tilde{u}(\imath'X)$ and $\tilde{F}\imath'X = \imath'FX$. This shows that $M$ is an invariant submanifold of $\mathbf{S}^{n+p-1}$ and $u$ is the restriction of the contact form $\tilde{u}$ of $\mathbf{S}^{n+p-1}$. □

Further, in the case when the ambient manifold is a complex projective space $\mathbf{P}^{\frac{n+p}{2}}(\mathbf{C})$, the theorem to be proved is the following

**Theorem 21.3.** [25] *Let $M$ be a complete $n$-dimensional CR submanifold of maximal CR dimension of a complex projective space $\mathbf{P}^{\frac{n+p}{2}}(\mathbf{C})$ which satisfies the condition*

$$h(FX, Y) - h(X, FY) = g(FX, Y)\eta, \quad \eta \in T^{\perp}(M) \tag{21.39}$$

*for all $X$, $Y \in T(M)$, where $\eta$ does not have zero points. Then one of the following holds:*

(1)  *$M$ is congruent to a geodesic hypersphere $M_{0,k}^{C}$ for $k = \frac{n-1}{2}$;*

(2)  *$M$ is congruent to $M(n, \theta)$;*

(3)  *there exists a geodesic hypersphere $S$ of $\mathbf{P}^{\frac{n+p}{2}}(\mathbf{C})$ such that $M$ is an invariant submanifold by the almost contact structure $F'$ of $S$.*

We begin by proving

**Lemma 21.8.** *Let $M$ be an $n$-dimensional CR submanifold of maximal CR dimension of a complex projective space $\mathbf{P}^{\frac{n+p}{2}}(\mathbf{C})$ which satisfies the condition (21.1). Then the multiplicity of the eigenvalue $\alpha$ is one.*

*Proof.* First we consider the case that $\rho \neq 2\lambda$. Then, by Theorem 21.1, $M$ and $A$ are respectively regarded as a real hypersurface of $\mathbf{P}^{\frac{n+1}{2}}(\mathbf{C})$ and the shape operator of $M$ in $\mathbf{P}^{\frac{n+1}{2}}(\mathbf{C})$. Let $X \perp U$ be an eigenvector of $A$ corresponding to the eigenvalue $\alpha$. Then from (21.2), we obtain that $FX$ is an eigenvector of $A$ corresponding to $\rho - \alpha$. Since, using Lemma 21.9, it follows that $A$ has at most three distinct eigenvalues, we conclude $\rho - \alpha = \alpha$ or $\rho - \alpha = \lambda$, where $\lambda$ is the solution of (21.35). Therefore, let $\rho - \alpha = \lambda$. Then, substituting this into (21.35) we obtain $2\alpha^2 - \rho\alpha + 2 = 0$. Using (21.35) again, we get $\lambda = \alpha$ or $\lambda = \rho - \alpha$. Thus solutions of (21.35) are $\alpha$ and $\rho - \alpha$. This means that $A$ has only two distinct eigenvalues. Since $A$ is regarded as the shape operator of a real hypersurface in $\mathbf{P}^{\frac{n+1}{2}}(\mathbf{C})$, we can apply Lemma 19.8, more precisely Remark 19.1, and obtain that the multiplicity of $\alpha$ is 1, which is a contradiction.

Next we consider the case $\rho = 2\lambda$, where $X$ is an eigenvector of $A$ with corresponding eigenvalue $\lambda \neq \alpha$. Let the multiplicity of $\alpha$ be $r$. Then

$$\text{trace } A = r\alpha + (n - r)\frac{\rho}{2}. \tag{21.40}$$

Using (21.2) and (15.6) we get

$$-AX + u(AX)U + FAFX = \rho(-X + u(X)U). \tag{21.41}$$

Taking the trace of (21.41), we obtain

$$\text{trace } A = \frac{\rho}{2}(n-1) + \alpha. \tag{21.42}$$

Using (21.40) and (21.42), we get

$$(r-1)(\alpha - \frac{\rho}{2}) = 0.$$

Since $\alpha \neq \frac{\rho}{2}$, it follows $r = 1$. □

**Proof of Theorem 21.3.** Lemma 21.8 shows that as a real hypersurface of $\mathbf{P}^{\frac{n+1}{2}}(\mathbf{C})$, $M$ has at most three distinct principal curvatures and the multiplicity of the eigenvalue $\alpha$ is 1.

If $M$ has only two distinct principal curvatures, by Theorem 19.3, $M$ is a geodesic hypersphere $M_{0,k}^C$ for $k = \frac{n-1}{2}$.

If $M$ has three distinct principal curvatures $\lambda_1$, $\lambda_2$ and $\alpha$, by Lemma 16.3, as a hypersurface of $\mathbf{S}^{n+2}$, $\pi^{-1}(M)$ has four principal curvatures $\lambda_1$, $\lambda_2$, $\mu$ and $-\frac{1}{\mu}$ with respective multiplicities $\frac{n-1}{2}$, $\frac{n-1}{2}$, 1 and 1. Thus, by Remark 12.1, $\pi^{-1}(M)$ is congruent to $M'(n,\theta)$ and $M = \pi M'(n,\theta) = M(n,\theta)$.

Hence, when $\rho \neq 2\lambda$, $M$ is congruent to $M_{k,0}^C$ for $k = \frac{n-1}{2}$ or to $M(n,\theta)$.

Next we consider the case $\rho = 2\lambda$. Let $X$ be an eigenvector of $A$ with corresponding eigenvalue $\lambda \neq \alpha$. Using (21.2), we conclude that

$$AFX = (\rho - \lambda)FX,$$

namely, it follows that $FX$ is an eigenvector with the corresponding eigenvalue $\rho - \lambda = 2\lambda - \lambda = \lambda$. Therefore, the only eigenvalues of $A$ are $\alpha$ and $\lambda = \frac{\rho}{2}$. Hence, from Theorem 19.2, it follows that there exists a geodesic hypersphere $S$ of $\mathbf{P}^{\frac{n+p}{2}}(\mathbf{C})$ such that $M$ lies on $S$.

We can now proceed analogously to the proof of Theorem 21.2 and conclude that $M$ is $F'$ invariant, which completes the proof of Theorem 21.3. □

*Remark* 21.3. The case when the ambient manifold $\overline{M}$ is a complex hyperbolic space $\mathbf{H}^{\frac{n+p}{2}}(\mathbf{C})$ is studied and the complete classification is given in [26] while the main results appear in [22].

# Invariant submanifolds of real hypersurfaces of complex space forms

In Remark 15.1 we recalled that real hypersurfaces of a complex manifold admit a naturally induced almost contact structure $F'$ from the almost complex structure of the ambient manifold. In Theorem 21.3 we proved that if $M$ is a complete $n$-dimensional CR submanifold of maximal CR dimension of a complex projective space $\mathbf{P}^{\frac{n+p}{2}}(\mathbf{C})$ satisfying the condition (21.1), then $M$ is congruent to a geodesic hypersphere $M_{0,k}^C$ for $k = \frac{n-1}{2}$, or to $M(n,\theta)$, or there exists a geodesic hypersphere $S$ of $\mathbf{P}^{\frac{n+p}{2}}(\mathbf{C})$ such that $M$ is an invariant submanifold by the almost contact structure $F'$ of $S$. It is easy to check that for the geodesic hypersphere $M_{0,k}^C$ for $k = \frac{n-1}{2}$, the following relation is satisfied:

$$A'F' + F'A' = \rho F', \tag{22.1}$$

where $A'$ is its shape operator. An easy computation, using Lemma 3.4 in [56], shows that $M(n,\theta)$ also satisfies relation (22.1).

Consequently, it appears interesting to solve the following problem

Does an $F'$-invariant submanifold of a geodesic hypersphere $S$ of a complex space form satisfy the condition (21.1)?

First, we prove that any odd-dimensional $F'$-invariant submanifold of a real hypersurface of a complex manifold $\overline{M}$ is a CR submanifold of maximal CR dimension. Considering this problem, it is natural to continue exploring Example 15.3 with more details.

Let $\overline{M}$ be a real $(m+1)$-dimensional complex manifold with natural almost complex structure $J$ and a Hermitian metric $\overline{g}$.

We consider a real hypersurface $M'$ of $\overline{M}$ and an $n$-dimensional submanifold $M$ of $M'$ with immersions $\imath_1$ and $\imath_0$, respectively. Then $M$ is a submanifold of $\overline{M}$ with the immersion $\imath = \imath_1 \imath_0$. The Riemannian metric $g'$ of $M'$ and $g$ of $M$ are induced from the Hermitian metric $\overline{g}$ of $\overline{M}$ in such a way that

M. Djorić, M. Okumura, *CR Submanifolds of Complex Projective Space*, 151
Developments in Mathematics 19, DOI 10.1007/978-1-4419-0434-8_22,
© Springer Science+Business Media, LLC 2010

$$g'(X',Y') = \overline{g}(\imath_1 X', \imath_1 Y'), \quad \text{for} \quad X', Y' \in T(M'),$$
$$g(X,Y) = g'(\imath_0 X, \imath_0 Y) = \overline{g}(\imath X, \imath Y), \quad \text{for} \quad X, Y \in T(M).$$

Let us denote by $\xi$ the unit normal local field to $M'$ in $T(\overline{M})$. Since a real hypersurface is a CR submanifold of maximal CR dimension (see Example 15.1), using the results of Section 15, it follows that $M'$ is endowed with the induced almost contact structure $(F', u', U')$ which satisfies (15.6), (15.7), (15.9) and (15.27).

Now, let us assume that $M$ is invariant under the action of the almost contact tensor $F'$, that is, $F'T(M) \subset T(M)$, for the tangent bundle $T(M)$. Consequently, if we denote by $U^\perp$ the normal part of $U'$ in $M'$, we may write

$$F'\imath_0 X = \imath_0 F X, \qquad U' = \imath_0 U + U^\perp, \tag{22.2}$$

where $F$ is an endomorphism acting on $T(M)$ and we can deduce

**Proposition 22.1.** *For an $F'$-invariant submanifold $M$ of $M'$, only the following two cases for the vector field $U'$ can occur:*

(1) *$U'$ is always tangent to $M$ and $M$ is necessarily odd-dimensional.*

(2) *$U'$ is never tangent to $M$ and $M$ is necessarily even-dimensional.*

*Proof.* From (15.6) and (22.2), it follows

$$F'^2 \imath_0 X = -\imath_0 X + u'(\imath_0 X)\imath_0 U + u'(\imath_0 X)U^\perp,$$
$$F'^2 \imath_0 X = F'\imath_0 F X = \imath_0 F^2 X.$$

Comparing the tangential part and normal part of the above equations, we conclude

$$F^2 X = -X + u'(\imath_0 X)U, \qquad u'(\imath_0 X)U^\perp = 0. \tag{22.3}$$

If $u'(\imath_0 X) = 0$ is satisfied at a point $x$ of $M$, the first equation of (22.3) reads $F^2 X = -X$. Thus, $F$ is an almost complex structure and $M$ is even-dimensional. Moreover, from

$$0 = u'(\imath_0 X) = g'(U', \imath_0 X) = g'(\imath_0 U + U^\perp, \imath_0 X) = g(U, X),$$

for all $X \in T(M)$, we conclude $U = 0$. Therefore, using (22.2), it follows $U' = U^\perp$, which shows that $U'$ is never tangent to $M$.

If $u'(\imath_0 X) \neq 0$ at some point $x \in M$, then it follows from (22.3) that $U^\perp = 0$ at $x \in M$, that is, $U' = \imath_0 U$. This shows that $U'$ is always tangent to $M$ and from (22.3) we conclude

$$F^2 X = -X + g(U, X)U.$$

Using the notation $\mathcal{D}_x = \{X \in T_x(M) : g(X, U) = 0\}$, we conclude that $F$ acts as an almost complex structure on $\mathcal{D}_x$ and therefore, $\mathcal{D}_x$ is even-dimensional. Consequently,

$$T_x(M) = \mathcal{D}_x \oplus \mathrm{span}\{U\}$$

and $M$ is odd-dimensional, which completes the proof.    □

For an odd-dimensional submanifold $M$, using $U' = \imath_0 U$, we compute

$$1 = g'(U', U') = g'(\imath_0 U, \imath_0 U) = g(U, U),$$
$$F^2 X = -X + u'(\imath_0 X)U = -X + g'(U', \imath_0 X)U = -X + g(U, X)U,$$
$$0 = F'U' = F'\imath_0 U = \imath_0 FU,$$

and, consequently, $FU = 0$. Hence, if we define the one-form $u$ on $M$ by $u(X) = g(U, X)$, then $u(X) = u'(\imath_0 X)$ and $(F, u, U)$ defines an almost contact structure on $M$.

Further, let $\xi'_1, \ldots, \xi'_p$ $(p = m - n)$ denotes the orthonormal normal frame field to $M$ in $M'$. Since

$$g'(F'\xi'_a, \imath_0 X) = -g'(\xi'_a, F'\imath_0 X) = -g'(\xi'_a, \imath_0 FX) = 0,$$

we conclude that the normal space $\mathrm{span}\{\xi'_1, \ldots, \xi'_p\}_x$ is an $F'$-invariant subspace of $T_x(M')$ at each point $x \in M$. Hence we may write

$$F'\xi'_a = \sum_{b=1}^{p} P_{ab}\xi'_b.$$

Now, we consider the submanifold $M$ of $M'$ as a submanifold of a complex manifold $\overline{M}$. We choose an orthonormal normal frame $\xi_1, \ldots, \xi_{p+1}$ of $M$ in $\overline{M}$ in such a way that

$$\xi_{p+1} = \xi, \qquad \xi_a = \imath_1 \xi'_a, \qquad a = 1, \ldots, p.$$

Then, from the above discussions, we easily conclude

$$J\imath X = \imath FX + u(X)\xi, \tag{22.4}$$
$$J\xi = -\imath U, \tag{22.5}$$
$$J\xi_a = J\imath_1 \xi'_a = \imath_1 F'\xi'_a = \sum_{b=1}^{p} P_{ab}\xi_b, \quad a = 1, \ldots, p. \tag{22.6}$$

This shows that $\mathrm{span}\{\xi_1, \ldots, \xi_p\}_x$ is a $J$-invariant subspace of $T_x(\overline{M})$.

Denoting by $\nabla$ and $\nabla'$ the Riemannian connections of $M$ and $M'$, respectively, the Gauss formula (5.1) yields

$$\nabla'_X \imath_0 Y = \imath_0 \nabla_X Y + \sum_{a=1}^{p} g(A'_a X, Y)\xi'_a, \tag{22.7}$$

where $A'_a$'s are the shape operators of $M$ with respect to $\xi'_a$. We denote by $A$ and $A_a$ $(a = 1, \ldots, p)$, the shape operators of $M$ with respect to $\xi$ and $\xi_a$, respectively. Then,

$$\overline{\nabla}_X \imath Y = \imath \nabla_X Y + g(AX, Y)\xi + \sum_{a=1}^{p} g(A_a X, Y)\xi_a, \tag{22.8}$$

where $\overline{\nabla}$ is the Riemannian connection of $\overline{M}$. On the other hand, using (5.1) and (22.7), we compute

$$\overline{\nabla}_X \imath Y = \overline{\nabla}_X \imath_1 \imath_0 Y = \imath_1 \nabla'_X \imath_0 Y + g'(A' \imath_0 X, \imath_0 Y)\xi$$

$$= \imath_1 (\imath_0 \nabla_X Y + \sum_{a=1}^{p} g(A'_a X, Y)\xi'_a) + g'(A' \imath_0 X, \imath_0 Y)\xi$$

$$= \imath \nabla_X Y + g'(A' \imath_0 X, \imath_0 Y)\xi + \sum_{a=1}^{p} g(A'_a X, Y)\xi_a. \tag{22.9}$$

Comparing (22.8) and (22.9), we conclude

$$g(AX, Y) = g'(A' \imath_0 X, \imath_0 Y), \qquad g(A_a X, Y) = g(A'_a X, Y), \tag{22.10}$$

from which it follows $A_a = A'_a$ for $a = 1, \ldots, p$.

The Weingarten formulae (5.6) for $M$ are

$$\overline{\nabla}_X \xi = -\imath AX + \sum_{a=1}^{p} s_a(X)\xi_a$$

$$= \imath_1 (-\imath_0 AX + \sum_{a=1}^{p} s_a(X)\xi'_a) \tag{22.11}$$

$$\overline{\nabla}_X \xi_a = -\imath A_a X - s_a(X)\xi + \sum_{b=1}^{p} s_{ab}(X)\xi_b, \tag{22.12}$$

where $s_a = s_{p+1\,a}$, $s_{ab}$ are the components of the third fundamental form of $M$ in $\overline{M}$ and $a = 1, \ldots, p$. On the other hand,

$$\overline{\nabla}_X \xi = -\imath_1 A' \imath_0 X, \tag{22.13}$$

$$\overline{\nabla}_X \xi_a = \imath_1 \nabla'_X \xi'_a + g'(A' \imath_0 X, \xi'_a)\xi$$

$$= -\imath A'_a X + \sum_{b=1}^{p} s'_{ab}(X)\xi_b + g'(A' \imath_0 X, \xi'_a)\xi, \tag{22.14}$$

where we denote by $s'_{ab}$ the components of the third fundamental form of $M$ in $M'$ and $a = 1, \ldots, p$. Comparing the tangential part and normal part of the equations (22.11)–(22.14), we obtain

$$A' \imath_0 X = \imath_0 AX - \sum_{a=1}^{p} s_a(X)\xi'_a,$$

$$s_a(X) = -g'(A' \imath_0 X, \xi'_a), \qquad s'_{ab} = s_{ab}. \tag{22.15}$$

Now, let $\overline{M}$ be a Kähler manifold. Then differentiating (22.5) covariantly and using (22.6), (22.8), (22.11) and (22.4), we compute

$$J\overline{\nabla}_X\xi = -\overline{\nabla}_X\imath U$$

$$= -\imath\nabla_X U - g(AX,U)\xi - \sum_{a=1}^{p} g(A_aX,U)\xi_a,$$

$$J\overline{\nabla}_X\xi = J(-\imath AX + \sum_{a=1}^{p} s_a(X)\xi_a)$$

$$= -\imath FAX - u(AX)\xi + \sum_{a,b=1}^{p} s_a(X)P_{ab}\xi_b$$

and consequently it follows

$$\nabla_X U = FAX, \qquad g(A_aX,U) = -\sum_{b=1}^{p} s_b(X)P_{ba}. \tag{22.16}$$

Let $\overline{R}$ and $R'$ denote the curvature tensors of $\overline{M}$ and its hypersurface $M'$, respectively. Then the Gauss equation (5.22) and the Codazzi equation (5.23) for the normals $\xi_a$, $a = 1,\ldots,p$ of $M$ in $\overline{M}$ yield

$$\overline{g}(\overline{R}(\imath_1 X', \imath_1 Y')\imath_1 Z', \imath_1 W') = g'(R'(X',Y')Z', W') - g'(A'Y', Z')g'(A'X', W')$$
$$+ g'(A'X', Z')g'(A'Y', W'), \tag{22.17}$$

$$\overline{g}(\overline{R}(\imath X, \imath Y)\imath Z, \xi_a) = g((\nabla_X A_a)Y - (\nabla_Y A_a)X, Z)$$
$$+ s_a(X)g(AY, Z) - s_a(Y)g(AX, Z) \tag{22.18}$$
$$+ \sum_{b=1}^{p}\{s_{ba}(X)g(A_bY, Z) - s_{ba}(Y)g(A_bX, Z)\}.$$

Now, let the ambient manifold $\overline{M}$ be a complex space form. Then its curvature tensor $\overline{R}$ is given by (9.21), for some constant $k$ and therefore

$$\overline{g}(\overline{R}(\imath X, \imath Y)\imath Z, \xi_a) = 0, \tag{22.19}$$

since $\text{span}\{\xi_1,\ldots,\xi_p\}$ is $J$-invariant. Consequently, using (22.18) and (22.19), it follows

$$(\nabla_X A_a)Y - (\nabla_Y A_a)X = s_a(Y)AX - s_a(X)AY$$
$$+ \sum_{b=1}^{p}\{s_{ab}(X)A_bY - s_{ab}(Y)A_bX\}. \tag{22.20}$$

Now we consider the case when the shape operator $A'$ of the real hypersurface $M'$ has the form

$$A'X' = \alpha X' + \beta u'(X')U', \tag{22.21}$$

and we prove the following

**Theorem 22.1.** *Let $M'$ be a real hypersurface of a nonflat complex space form whose shape operator $A'$ has the form (22.21) and let $M$ be an $F'$-invariant submanifold of $M'$. Then $U'$ is always tangent to $M$ and $M$ is odd-dimensional.*

*Proof.* Suppose, contrary to our claim, that $U'$ is not tangent to $M$. Then, using the second equation of (22.3), we compute

$$0 = u'(\imath_0 X) = g'(U', \imath_0 X) = g'(\imath_0 U + U^\perp, \imath_0 X) = g(U, X),$$

for all $X \in T(M)$ and we conclude $U = 0$ and $U' = U^\perp$. Therefore, let us choose the other normals in such a way that $\xi_1' = U'$ and that $A_{U'}$ denotes the shape operator for the normal $U'$. Since the shape operator $A'$ has the form given by (22.21), using relations (15.7) and (15.27), we compute

$$\nabla_X' U' = F'A'\imath_0 X = F'(\alpha \imath_0 X + \beta u'(\imath_0 X)U') = \alpha F'\imath_0 X = \alpha \imath_0 FX.$$

On the other hand, using the Weingarten formula (5.6) for the normal $U'$, we have

$$\nabla_X' U' = -\imath_0 A_{U'} X + \sum_{a=2}^{p} s_{Ua}(X)\xi_a'.$$

Comparing the above two equations, we conclude $A_{U'} X = -\alpha FX$. Since $A_{U'}$ is symmetric and $F$ is skew-symmetric, it follows $\alpha = 0$ and consequently, using (15.7), (15.27), we compute

$$A'X' = \beta u'(X')U', \qquad \nabla_X' U' = 0. \tag{22.22}$$

Since $\overline{M}$ is a complex space form, using (9.21), (15.2), (22.17) and (22.22), we obtain

$$R'(X', Y')Z' = k\{g'(Y', Z')X' - g'(X', Z')Y' + g'(F'Y', Z')F'X' \\ - g'(F'X', Z')F'Y' - 2g'(F'X', Y')F'Z'\}$$

and consequently

$$R'(X', Y')U' = k\{g'(Y', U')X' - g'(X', U')Y'\}. \tag{22.23}$$

On the other hand, the second equation of (22.22) implies

$$R'(X', Y')U' = 0.$$

Consequently, using (22.23), it follows $k = 0$, which is a contradiction, since we have assumed that the ambient manifold $\overline{M}$ is nonflat. Therefore, $U'$ is tangent to $M$. Repeating the same procedure as at the end of the proof of Proposition 22.1, we conclude that $M$ is odd-dimensional, which completes the proof.    □

*Remark* 22.1. An obvious question to ask is

*Does the assertion of Theorem 22.1 continue to hold when the ambient manifold $\overline{M}$ is a complex Euclidean space?*

We construct here a counterexample.

*Example* 22.1. Let $M'$ be a real hypersurface of $\mathbf{C}^{n+1}$ defined by

$$M' = \mathbf{C}^n \times S^1 = \{(z^1, \ldots, z^n, e^{\sqrt{-1}\theta}) \in \mathbf{C}^{n+1}\}$$
$$= \{(x^1, y^1, \ldots, x^n, y^n, \cos\theta, \sin\theta) \in \mathbf{E}^{2n+2}\},$$

where $z^i = x^i + \sqrt{-1}y^i$, $i = 1, \ldots, n$. We note that the defining function $f$ of $M'$ is

$$f(x^1, y^1, \ldots, x^n, y^n, x^{n+1}, y^{n+1}) = (x^{n+1})^2 + (y^{n+1})^2 - 1. \qquad (22.24)$$

With respect to local coordinates $(u^1, \ldots, u^{2n}, \theta)$ of $M'$, we compute

$$\frac{\partial x^i}{\partial u^{2j-1}} = \frac{\partial y^i}{\partial u^{2j}} = \delta^i_j, \quad j, i = 1, \ldots, n$$

$$\frac{\partial x^i}{\partial u^{2j}} = \frac{\partial y^i}{\partial u^{2j-1}} = 0, \quad j, i = 1, \ldots, n$$

$$\frac{\partial x^i}{\partial \theta} = \frac{\partial y^i}{\partial \theta} = 0, \quad i = 1, \ldots, n$$

$$\frac{\partial x^{n+1}}{\partial u^{2j-1}} = \frac{\partial x^{n+1}}{\partial u^{2j}} = \frac{\partial y^{n+1}}{\partial u^{2j-1}} = \frac{\partial y^{n+1}}{\partial u^{2j}} = 0, \quad j = 1, \ldots, n$$

$$\frac{\partial x^{n+1}}{\partial \theta} = -\sin\theta, \quad \frac{\partial y^{n+1}}{\partial \theta} = \cos\theta. \qquad (22.25)$$

For the immersion $\imath_1 : M' \to \mathbf{E}^{2n+2}$, we have

$$\imath_1\left(\frac{\partial}{\partial u^{2j-1}}\right) = \sum_{k=1}^{n+1}\left(\frac{\partial x^k}{\partial u^{2j-1}}\frac{\partial}{\partial x^k} + \frac{\partial y^k}{\partial u^{2j-1}}\frac{\partial}{\partial y^k}\right) = \frac{\partial}{\partial x^j}, \qquad (22.26)$$

$$\imath_1\left(\frac{\partial}{\partial u^{2j}}\right) = \frac{\partial}{\partial y^j}, \qquad (22.27)$$

$$\imath_1\left(\frac{\partial}{\partial \theta}\right) = -\sin\theta\frac{\partial}{\partial x^{n+1}} + \cos\theta\frac{\partial}{\partial y^{n+1}}. \qquad (22.28)$$

Further, using the defining function (22.24), we compute

$$\operatorname{grad} f = \sum_{i=1}^{n+1}\left(\frac{\partial f}{\partial x^i}\frac{\partial}{\partial x^i} + \frac{\partial f}{\partial y^i}\frac{\partial}{\partial y^i}\right)$$

$$= 2\left(x^{n+1}\frac{\partial}{\partial x^{n+1}} + y^{n+1}\frac{\partial}{\partial y^{n+1}}\right)$$

and consequently $|\mathrm{grad}\, f|^2 = 4$. Hence the unit normal vector field $\xi$ is given by

$$\xi = x^{n+1}\frac{\partial}{\partial x^{n+1}} + y^{n+1}\frac{\partial}{\partial y^{n+1}} = \cos\theta\frac{\partial}{\partial x^{n+1}} + \sin\theta\frac{\partial}{\partial y^{n+1}} \qquad (22.29)$$

and therefore

$$J\xi = -\sin\theta\frac{\partial}{\partial x^{n+1}} + \cos\theta\frac{\partial}{\partial y^{n+1}} = -\imath_1 U'. \qquad (22.30)$$

Comparing (22.28) and (22.30), we deduce that

$$U' = -\frac{\partial}{\partial\theta}. \qquad (22.31)$$

Now, let $M$ be a submanifold of $M'$ with the immersion $\imath_0$ defined by

$$M = \{(x^1, y^1, \ldots, x^r, y^r, 0, \ldots, 0, \cos\theta, \sin\theta)|\, \theta = const.\}.$$

For a local coordinate system $(v^1, \ldots, v^{2r})$ of $M$, we have

$$x^i = u^{2i-1} = v^{2i-1}, \quad y^i = u^{2i} = v^{2i}, \quad i = 1, \ldots, 2r$$

and

$$\imath_0\left(\frac{\partial}{\partial v^{2i-1}}\right) = \frac{\partial}{\partial u^{2i-1}}, \quad \imath_0\left(\frac{\partial}{\partial v^{2i}}\right) = \frac{\partial}{\partial u^{2i}}. \qquad (22.32)$$

Let $g'$ denote the induced metric from the Euclidean metric $\langle,\rangle$ of the ambient manifold $\mathbf{E}^{2n}$. Then, for $i = 1, \ldots, 2r$, using (22.26), (22.28), (22.31), (22.32) we compute

$$g'\left(\imath_0\left(\frac{\partial}{\partial v^{2i-1}}\right), U'\right) = g'\left(\imath_0\left(\frac{\partial}{\partial v^{2i-1}}\right), -\frac{\partial}{\partial\theta}\right)$$

$$= -\left\langle\imath_1\left(\frac{\partial}{\partial u^{2i-1}}\right), \imath_1\left(\frac{\partial}{\partial\theta}\right)\right\rangle$$

$$= -\left\langle\frac{\partial}{\partial x^i}, -\sin\theta\frac{\partial}{\partial x^{n+1}} + \cos\theta\frac{\partial}{\partial y^{n+1}}\right\rangle = 0.$$

By a similar argument, it follows

$$g'\left(\imath_0\left(\frac{\partial}{\partial v^{2i}}\right), U'\right) = 0$$

and we conclude that $U'$ is normal to $M$.

Now, using relation (22.29) we compute

$$\nabla^E_{\frac{\partial}{\partial u^i}}\xi = \nabla^E_{\frac{\partial}{\partial v^i}}\xi = 0, \qquad (22.33)$$

$$\nabla^E_{\frac{\partial}{\partial\theta}}\xi = \imath_1\left(\frac{\partial}{\partial\theta}\right) = -\imath_1 U', \qquad (22.34)$$

where $\nabla^E$ is the covariant derivative in $\mathbf{C}^{n+1}$. On the other hand, for the hypersurface $M'$, the Weingarten formula (5.6) implies

$$\nabla^E_{X'}\xi = -\imath_1 A'X', \tag{22.35}$$

where $A'$ is the shape operator with respect to the normal vector field $\xi$. Using (22.33), (22.34) and (22.35), it follows

$$A'U' = -U', \tag{22.36}$$
$$A'X' = 0, \quad \text{for} \quad X' \perp U'. \tag{22.37}$$

Therefore, decomposing $X'$ as $X' = Y' + aU'$, where $Y' \perp U'$, using (22.36) and (22.37), we compute $A'X' = -aU'$ and $a = -g'(A'X', U')$. Consequently, the shape operator $A'$ has the form (22.21), namely,

$$A'X' = -g'(U', X')U'.$$

In order to prove that $M$ is $F'$-invariant, using (22.26), (22.27), (22.32), we note the following:

$$\imath\left(\frac{\partial}{\partial v^{2i-1}}\right) = \imath_1\imath_0\left(\frac{\partial}{\partial v^{2i-1}}\right) = \imath_1\left(\frac{\partial}{\partial u^{2i-1}}\right) = \frac{\partial}{\partial x^i},$$
$$\imath\left(\frac{\partial}{\partial v^{2i}}\right) = \imath_1\imath_0\left(\frac{\partial}{\partial v^{2i}}\right) = \imath_1\left(\frac{\partial}{\partial u^{2i}}\right) = \frac{\partial}{\partial y^i}.$$

On the other hand, since $U'$ is normal to $M$, using (22.27) and (22.32), we obtain

$$\frac{\partial}{\partial y^i} = J\frac{\partial}{\partial x^i} = J\imath_1\imath_0\left(\frac{\partial}{\partial v^{2i-1}}\right)$$
$$= \imath_1 F'\imath_0\left(\frac{\partial}{\partial v^{2i-1}}\right) + g'\left(\frac{\partial}{\partial v^{2i-1}}, U'\right)\xi$$
$$= \imath_1 F'\imath_0\left(\frac{\partial}{\partial v^{2i-1}}\right). \tag{22.38}$$

Using (22.27) and (22.32) it follows

$$\frac{\partial}{\partial y^i} = \imath_1\left(\frac{\partial}{\partial u^{2i}}\right) = \imath_1\imath_0\left(\frac{\partial}{\partial v^{2i}}\right). \tag{22.39}$$

Considering relations (22.38) and (22.39), we deduce

$$\imath_1 F'\imath_0\left(\frac{\partial}{\partial v^{2i-1}}\right) = \imath_1\imath_0\left(\frac{\partial}{\partial v^{2i}}\right)$$

from which we conclude

$$F'\imath_0 \left( \frac{\partial}{\partial v^{2i-1}} \right) = \imath_0 \left( \frac{\partial}{\partial v^{2i}} \right). \tag{22.40}$$

A slight change in the above proof shows

$$F'\imath_0 \left( \frac{\partial}{\partial v^{2i}} \right) = -\imath_0 \left( \frac{\partial}{\partial v^{2i-1}} \right). \tag{22.41}$$

Combining (22.40) and (22.41) we have proved that $M$ is an $F'$-invariant submanifold of $M'$.  ◇

Now we consider the case when $M$ is an odd-dimensional submanifold of a real hypersurface $M'$ whose shape operator $A'$ satisfies relation (20.14). Using (22.10) and the first equation of (22.16), we obtain

$$AX = \alpha X + \beta u(X)U, \qquad \nabla_X U = \alpha FX \tag{22.42}$$

and we prove

**Lemma 22.1.** *If the shape operator $A'$ of the real hypersurface $M'$ satisfies relation (22.21), it follows $s_a = 0$ and $A_a U = 0$, $a = 1, \ldots, p$, for its $F'$-invariant submanifold $M$.*

*Proof.* Using relations (22.21) and (22.15), we obtain

$$s_a(X) = -g'(A'\imath_0 X, \xi'_a) = -g'(\alpha\imath_0 X + \beta u'(\imath_0 X)\imath_0 U, \xi'_a) = 0.$$

Since $A_a$ is symmetric, using the second equation of (22.16), we compute $g(A_a X, U) = g(A_a U, X) = 0$ for any $X \in T(M)$, which completes the proof.  □

Further, differentiating relation $A_a U = 0$ covariantly and using (22.42), we obtain $(\nabla_X A_a)U + \alpha A_a FX = 0$. Hence, we have

$$g((\nabla_X A_a)Y - (\nabla_Y A_a)X, U) = g((\nabla_X A_a)U, Y) - g((\nabla_Y A_a)U, X)$$
$$= -\alpha\{g(A_a FX, Y) - g(A_a FY, X)\}. \tag{22.43}$$

On the other hand, using relations (22.20) and (5.9) and Lemma 22.1, it follows

$$g((\nabla_X A_a)Y - (\nabla_Y A_a)X, U) = \sum_{b=1}^{p} \{s_{ab}(X)g(A_b U, Y) - s_{ab}(Y)g(A_b U, X)\} = 0,$$

which together with relation (22.43) gives $\alpha g((A_a F + F A_a)X, Y) = 0$. Therefore, for $\alpha \neq 0$, we conclude

$$A_a F + F A_a = 0. \tag{22.44}$$

Finally, for $\alpha \neq 0$, using the relation (5.7) between the second fundamental form and the shape operator and relations (22.42), (22.44), we calculate

$$
\begin{aligned}
h(FX, Y) - h(X, FY) &= \{g(AFX, Y) - g(AFY, X)\}\xi \\
&\quad + \sum_{a=1}^{p} \{g(A_a FX, Y) - g(A_a FY, X)\}\xi_a \\
&= 2\alpha g(FX, Y)\xi + \sum_{a=1}^{p} g((A_a F + FA_a)X, Y)\xi_a \\
&= 2\alpha g(FX, Y)\xi.
\end{aligned}
$$

In a complex projective space, the real hypersurface whose shape operator satisfies (22.21) is a geodesic hypersphere (see Theorem 19.3) and in this case $\alpha = \cot\theta$, $0 < \theta < \frac{\pi}{2}$. Therefore $\alpha \neq 0$ and the following theorem is established from the above discussion.

**Theorem 22.2.** *[27] Let $M'$ be a real hypersurface of a complex projective space whose shape operator $A'$ has the form (22.21). Then for any $F'$-invariant submanifold $M$ of $M'$, its second fundamental form $h$ satisfies the condition (21.1).*

Moreover, since the rank of $F$ is $n - 1$ and from the second equation of (22.42) it follows

$$
\begin{aligned}
du(X, Y) &= X(u(Y)) - Y(u(X)) - u([X, Y]) = g(\nabla_X U, Y) - g(\nabla_Y U, X) \\
&= \alpha(g(FX, Y) - g(FY, X)) = 2\alpha g(FX, Y),
\end{aligned}
$$

we conclude that $u \wedge (du)^k \neq 0$, $k = \frac{n-1}{2}$ and $u$ is a contact form.

**Corollary 22.1.** *Any $F'$-invariant submanifold $M$ of a real hypersurface $M'$ of a complex projective space whose shape operator $A'$ has the form (22.21), is a contact manifold.*

*Remark 22.2.* When $\overline{M}$ is a complex hyperbolic space $\mathbf{H}^{\frac{n+p}{2}}(\mathbf{C})$, the results analogous to those formulated as Theorem 22.2 are stated and proved in [27].

*Remark 22.3.* When the ambient manifold $\overline{M}$ is a complex projective space, the (22.21) on the shape operator $A'$ of the real hypersurface $M'$, which appears in Theorems 22.1 and 22.2, is equivalent to requiring that "the real hypersurface $M'$ has exactly two principal curvatures." This follows from Remark 19.1. However, if the ambient manifold is a complex hyperbolic space, it occurs that $M$ has exactly two principal curvatures, but the shape operator $A'$ fails to satisfy (20.14) (see [37] for more details).

# The scalar curvature of CR submanifolds of maximal CR dimension

In this section we first recall the so-called Bochner technique and we give a sufficient condition for a minimal CR submanifold $M^n$ of maximal CR dimension of the complex projective space $\mathbf{P}^{\frac{n+p}{2}}(\mathbf{C})$ to be $M_{r,s}^C$, $2r + 2s = n - 1$, namely, a tube over a totally geodesic complex subspace.

Since the ambient manifold is the complex projective space $\mathbf{P}^{\frac{n+p}{2}}(\mathbf{C})$ with Fubini-Study metric of constant holomorphic sectional curvature 4, using relation (15.28), we compute the Ricci tensor $Ric$ and the scalar curvature $\rho$ of $M$, respectively:

$$Ric(X,Y) = (n+2)g(X,Y) - 3u(X)u(Y) + (\text{trace } A)g(AX,Y) - g(A^2X,Y)$$
$$+ \sum_{a=1}^{q}\{(\text{trace } A_a)g(A_aX,Y) + (\text{trace } A_{a^*})g(A_{a^*}X,Y)$$
$$- g(A_a^2X,Y) - g(A_{a^*}^2X,Y)\}, \tag{23.1}$$

$$\rho = (n+3)(n-1) + (\text{trace } A)^2 - \text{trace } A^2$$
$$+ \sum_{a=1}^{q}\{(\text{trace } A_a)^2 + (\text{trace } A_{a^*})^2 - \text{trace } A_a^2 - \text{trace } A_{a^*}^2\}. \tag{23.2}$$

Now we prove the following

**Lemma 23.1.** *Let $M$ be an $n$-dimensional compact, minimal CR submanifold of maximal CR dimension of $\mathbf{P}^{\frac{n+p}{2}}(\mathbf{C})$. If the scalar curvature $\rho$ of $M$ satisfies*

$$\rho \geq (n+2)(n-1),$$

*then $F$ and $A$ commute, $A_a = A_{a^*} = 0$, $a = 1,\ldots,q$ and $\rho = (n+2)(n-1)$.*

*Proof.* The proof is based on the so-called *Bochner technique* (see [65]). Namely, using the famous Green's theorem, that is, the fact that on a compact manifold $M$,

M. Djorić, M. Okumura, *CR Submanifolds of Complex Projective Space*, Developments in Mathematics 19, DOI 10.1007/978-1-4419-0434-8_23, © Springer Science+Business Media, LLC 2010

$$\text{for} \quad \text{any} \quad X \in T(M), \quad \int_M \text{div} \, X * 1 = 0,$$

where $*1$ is the volume element of $M$, and calculating

$$\text{div} \, (\nabla_X X) - \text{div} \, ((\text{div} \, X)X),$$

K. Yano ([63], [64], [65]) established the following integral formula:

$$\int_M \{ \text{Ric}(X, X) + \frac{1}{2}|L(X)g|^2 - |\nabla X|^2 - (\text{div} \, X)^2 \} * 1 = 0, \qquad (23.3)$$

where $X$ is an arbitrary tangent vector field on $M$, $|Y|$ is the length of $Y$ with respect to the Riemannian metric $g$ of $M$ and $L(X)$ is the operator of Lie derivative with respect to $X$.

We put $X = U$ in (23.3) to obtain

$$\int_M \{ \text{Ric}(U, U) + \frac{1}{2}|L(U)g|^2 - |\nabla U|^2 - (\text{div} \, U)^2 \} * 1 = 0. \qquad (23.4)$$

On the other hand, making use of (15.27), we compute

$$\text{div} \, U = \text{trace} \, (FA) = 0, \qquad (23.5)$$

$$\sum_{a=1}^{q} |L(U)g|^2 = 2 \{ \text{trace} \, (FA)^2 + \text{trace} \, A^2 - g(A^2 U, U) \}, \qquad (23.6)$$

$$\sum_{a=1}^{q} |\nabla U|^2 = \text{trace} \, A^2 - g(A^2 U, U). \qquad (23.7)$$

Since $M$ is minimal submanifold, using Proposition 5.4, it follows

$$\text{trace} \, A = \text{trace} \, A_a = \text{trace} \, A_{a*} = 0, \quad a = 1, \ldots, q.$$

Therefore, using (23.1) and (23.2), we compute

$$\text{Ric}(U, U) = n - 1 - g(A^2 U, U) - \sum_{a=1}^{q'} \{ g(A_a^2 U, U) + g(A_{a*}^2 U, U) \}, \quad (23.8)$$

$$\rho = (n + 3)(n - 1) - \text{trace} \, A^2 - \sum_{a=1}^{q} \{ \text{trace} \, A_a^2 + \text{trace} \, A_{a*}^2 \}. \qquad (23.9)$$

Substituting (23.5), (23.7) and (23.8) into (23.4) and making use of (23.9), we conclude

$$\int_M \left\{ \frac{1}{2}|L(U)g|^2 + \rho - (n + 2)(n - 1) \right. \qquad (23.10)$$

$$\left. + \sum_{a=1}^{q} \text{trace} \, A_a^2 + \text{trace} \, A_{a*}^2 - g(A_a^2 U, U) - g(A_{a*}^2 U, U) \right\} * 1 = 0.$$

Now, we choose mutually orthonormal vector fields $e_1, \ldots, e_n$ in such a way that $e_n = U$. Since the shape operator is symmetric, it follows

$$\text{trace } A_a^2 + \text{trace } A_{a^*}^2 - g(A_a^2 U, U) - g(A_{a^*}^2 U, U)$$

$$= \sum_{i=1}^{n-1} \{g(A_a^2 e_i, e_i) + g(A_{a^*}^2 e_i, e_i)\}$$

$$= \sum_{i=1}^{n-1} \{g(A_a e_i, A_a e_i) + g(A_{a^*} e_i, A_{a^*} e_i)\} \geq 0.$$

Therefore, using relation (23.10), it follows that if $\rho \geq (n+2)(n-1)$, then the integrand is nonnegative and we compute

$$L(U)g = 0, \quad \rho = (n+2)(n-1), \quad A_a e_i = A_{a^*} e_i = 0, \qquad (23.11)$$

for $a = 1, \ldots, q$ and $i = 1, \ldots, n-1$. Consequently, it follows $FA = AF$, since using relation (15.27), we compute

$$0 = (L(U)g)(X, Y) = g(\nabla_X U, Y) + g(\nabla_Y U, X) = g((FA - AF)X, Y).$$

Moreover, using relation (23.11), it follows

$$A_a X = A_{a^*} X = 0, \quad \text{for any} \quad X \perp U, \qquad (23.12)$$

or equivalently

$$A_a FX = A_{a^*} FX = 0, \quad \text{for any} \quad X \in T(M). \qquad (23.13)$$

On the other hand, since $M$ is minimal, Proposition 5.4 and relation (15.20) imply

$$s_a(U) = s_{a^*}(U) = 0. \qquad (23.14)$$

Now, substituting $U$ instead of $Y$ in (15.21) and (15.22), and using relations (23.13) and (23.14), we obtain

$$g(FA_a X, U) = s_a(X), \quad g(FA_{a^*} X, U) = s_{a^*}(X). \qquad (23.15)$$

Since $F$ is skew-symmetric, using relation (15.7), we deduce from (23.15)

$$s_a(X) = 0 = s_{a^*}(X) = 0. \qquad (23.16)$$

Therefore, using relations (15.17) and (15.18), it follows $A_a U = A_{a^*} U = 0$, which together with (23.11) implies $A_a = A_{a^*} = 0$.

This completes the proof. $\qquad \square$

Now we prove the following:

**Theorem 23.1.** *Let $M$ be an $n$-dimensional compact, minimal CR subman-ifold of CR dimension $\frac{n-1}{2}$ in the complex projective space $\mathbf{P}^{\frac{n+p}{2}}(\mathbf{C})$. If the scalar curvature $\rho$ of $M$ is greater than or equal to $(n+2)(n-1)$, then there exists a totally geodesic complex projective subspace $\mathbf{P}^{\frac{n+1}{2}}(\mathbf{C})$ such that $M \subset \mathbf{P}^{\frac{n+1}{2}}(\mathbf{C})$.*

*Proof.* Define $N_0 = \{\xi \in T_x^\perp(M) : A_\xi = 0\}$. We note that, in this case,

$$N_0(x) = \operatorname{span}\{\xi_1(x), \ldots, \xi_q(x), \xi_{1^*}(x), \ldots, \xi_{q^*}(x)\}.$$

In fact, as a consequence of Lemma 23.1, $A_a = 0 = A_{a^*}$ for $a = 1, \ldots, q$, and therefore

$$\operatorname{span}\{\xi_1(x), \ldots, \xi_q(x), \xi_{1^*}(x), \ldots, \xi_{q^*}(x)\} \subset N_0(x).$$

On the other hand, for any $\eta \in N_0(x)$, we put $\eta = p^0\xi + \sum_{a=1}^q \{p^a \xi_a + p^{a^*} \xi_{a^*}\}$. Then

$$0 = A_\eta = p^0 A + \sum_{a=1}^q \{p^a A_a + p^{a^*} A_{a*}\} = p^0 A = 0,$$

since $A_a = A_{a^*} = 0$ for $a = 1, \ldots, q$. Hence $p^0 = 0$ and

$$\eta = \sum_{a=1}^q \{p^a \xi_a + p^{a^*} \xi_{a^*}\} \in \operatorname{span}\{\xi_1(x), \ldots, \xi_q(x), \xi_{1^*}(x), \ldots, \xi_{q^*}(x)\}.$$

Moreover, since $J\xi_a = \xi_{a^*}$, we have $JN_0(x) = N_0(x)$ and consequently

$$H_0(x) = JN_0(x) \cap N_0(x) = \operatorname{span}\{\xi_1(x), \ldots, \xi_q(x), \xi_{1^*}(x), \ldots, \xi_{q^*}(x)\}.$$

Hence the orthogonal complement $H_1(x)$ of $H_0(x)$ in $T_x^\perp(M)$ is span $\{\xi\}$.

Using relation (23.16), it follows that $H_1(x)$ is invariant under parallel translation with respect to the normal connection, and applying Theorem 14.3, we conclude the proof.    $\square$

From Theorem 23.1, we deduce that the submanifold $M$ can be regarded as a real hypersurface of $\mathbf{P}^{\frac{n+1}{2}}(\mathbf{C})$ which is a totally geodesic submanifold of $\mathbf{P}^{\frac{n+p}{2}}(\mathbf{C})$. In what follows we denote the totally geodesic submanifold $\mathbf{P}^{\frac{n+1}{2}}(\mathbf{C})$ by $M'$ and by $\imath_1$ the immersion of $M$ into $M'$ and by $\imath_2$ the totally geodesic immersion of $M'$ into $\mathbf{P}^{\frac{n+p}{2}}(\mathbf{C})$. Then, using the Gauss formula (5.1), it follows

$$\nabla'_X \imath_1 Y = \imath_1 \nabla_X Y + h'(X, Y) = \imath_1 \nabla_X Y + g'(A'X, Y)\xi', \qquad (23.17)$$

where $h'$ is the second fundamental form of $M$ in $M'$, $A'$ is the corresponding shape operator and $\xi'$ is the unit normal vector field to $M$ in $M'$. Since $\imath = \imath_2 \circ \imath_1$, we have

$$\overline{\nabla}_X \imath_2 \circ \imath_1 Y = \imath_2 \nabla'_X \imath_1 Y + \overline{h}(\imath_1 X, \imath_1 Y)$$
$$= \imath_2 (\imath_1 \nabla_X Y + g(A'X, Y)\xi'), \qquad (23.18)$$

since $M'$ is totally geodesic in $\mathbf{P}^{\frac{n+p}{2}}(\mathbf{C})$. Comparing relations (5.1) and (23.18), we conclude

$$\xi = \imath_2 \xi', \quad A = A'.$$

Further, as $M'$ is a complex submanifold of $\mathbf{P}^{\frac{n+p}{2}}(\mathbf{C})$, relation

$$J \imath_2 X' = \imath_2 J' X'$$

holds for any $X' \in T(M')$, where $J'$ is the induced complex structure of $M' = \mathbf{P}^{\frac{n+1}{2}}(\mathbf{C})$. Thus, using relation (15.2), we compute

$$J \imath X = J \imath_2 \circ \imath_1 X = \imath_2 J' \imath_1 X = \imath_2 (\imath_1 F'X + u'(X)\xi')$$
$$= \imath F'X + u'(X)\imath_2 \xi' = \imath F'X + u'(X)\xi. \qquad (23.19)$$

Comparing relations (23.19) and (15.2), we conclude

$$F = F', \quad u' = u.$$

Consequently, by Theorem 23.1, we deduce that $M$ is a real hypersurface of $\mathbf{P}^{\frac{n+1}{2}}(\mathbf{C})$ which satisfies $F'A' = A'F'$. Applying Theorem 16.3, we obtain

**Theorem 23.2.** [24] *If $M$ is an $n$-dimensional compact, minimal CR submanifold of maximal CR dimension of $\mathbf{P}^{\frac{n+p}{2}}(\mathbf{C})$, whose scalar curvature $\rho$ satisfies*

$$\rho \geq (n+2)(n-1),$$

*then $M$ is congruent to $M_{r,s}^C$ for some $r$, $s$ satisfying $2r + 2s = n - 1$.*

*Remark* 23.1. Theorem 23.2 was proved in [10] under the condition that the distinguished normal vector field $\xi$ is parallel with respect to the normal connection.

# References

[1] A. Bejancu, *CR-submanifolds of a Kähler manifold I*, Proc. Amer. Math. Soc. **69**, 135–142, (1978).

[2] A. Bejancu, *Geometry of CR-submanifolds*, D. Reidel Publ., Dordrecht, Holland, (1986).

[3] D. E. Blair, *Contact manifolds in Riemannian geometry*, Lecture Notes in Math., **509**, Springer, Berlin, (1976).

[4] D. E. Blair, *Riemannian geometry of contact and symplectic manifolds*, Progress in Mathematics, **203**, Birkhäuser, Boston, (2001).

[5] W. M. Boothby, *An introduction to differentiable manifolds and Riemannian geometry*, Academic Press, San Diego, (2003).

[6] E. Cartan, *Familles de surfaces isoparametriques dans les espaces à courbure constante*, Ann. Mat. Pura IV **17**, 177–191, (1938).

[7] E. Cartan, *Sur quelques familles remarquables d'hypersurfaces*, C. R. Congrés Math. Liége 30–41, (1939); Oeuvres completes Tome III, Vol. 2, 1481–1492.

[8] T. E. Cecil and P. J. Ryan, *Focal sets and real hypersurfaces in complex projective space*, Trans. Amer. Math. Soc. **269**, 481–499, (1982).

[9] B. Y. Chen, *Geometry of submanifolds*, Pure Appl. Math. **22**, Marcel Dekker, New York, (1973).

[10] Y. W. Choe and M. Okumura, *Scalar curvature of a certain CR-submanifold of complex projective space*, Arch. Math. **68**, 340–346, (1997).

[11] M. Djorić, *CR submanifolds of maximal CR dimension in complex projective space and its holomorphic sectional curvature*, Kragujevac J. Math. **25**, 171–178, (2003).

[12] M. Djorić, *Commutative condition on the second fundamental form of CR submanifolds of maximal CR dimension of a Kähler manifold*, "Complex, Contact and Symmetric Manifolds - In Honor of L.Vanhecke", Progress in Mathematics, **234**, Birkhäuser, Boston. Editors: O. Kowalski, E. Musso, D. Perrone, 105–120, (2005).

[13] M. Djorić, *Codimension reduction and second fundamental form of CR submanifolds in complex space forms*, J. Math. Anal. Appl. **356**, 237–241, (2009).

[14] M. Djorić and M. Okumura, *On contact submanifolds in complex projective space*, Math. Nachr. **202**, 17–28, (1999).

[15] M. Djorić and M. Okumura, *CR submanifolds of maximal CR dimension of complex projective space*, Arch. Math. **71**, 148–158, (1998).

[16] M. Djorić and M. Okumura, *CR submanifolds of maximal CR dimension in complex manifolds*, Proceedings of the Workshop PDE's, Submanifolds and Affine Differential Geometry, Banach center publications, Institute of Mathematics, Polish Academy of Sciences, Warsaw **57**, 89–99, (2002).

[17] M. Djorić and M. Okumura, *An application of an integral formula to CR submanifold of complex projective space*, Publ. Math. Debrecen **62**/1-2, 213–225, (2003).

[18] M. Djorić and M. Okumura, *On curvature of CR submanifolds of maximal CR dimension in complex projective space*, Izv. VuZov. Mat. No. 11 **498**, 15–23, (2003); Russ. Math. (Izv. VuZov) **47**, No. 11, 12–20, (2003).

[19] M. Djorić and M. Okumura, *Levi form of CR submanifolds of maximal CR dimension of complex space forms*, Acta Math. Hung. **102** (4), 297–304, (2004).

[20] M. Djorić and M. Okumura, *CR submanifolds of maximal CR dimension in complex space forms and second fundamental form*, Proceedings of the Workshop Contemporary Geometry and Related Topics, Belgrade, May 15-21, 2002, World Scientific, ISBN 981-238-432-4. Editors: N. Bokan, M. Djorić, A. T. Fomenko, Z. Rakić, J. Wess, 105–116, (2004).

[21] M. Djorić and M. Okumura, *Certain condition on the second fundamental form of CR submanifolds of maximal CR dimension of complex Euclidean space*, Ann. Global Anal. Geom. **30**, 383–396, (2006).

[22] M. Djorić and M. Okumura, *Certain contact submanifolds of complex space forms*, Proceedings of the Conference Contemporary Geometry and Related Topics, Belgrade, June 26-July 2, 2005. Editors: N. Bokan, M. Djorić, A. T. Fomenko, Z. Rakić, B. Wegner, J. Wess, 157–176, (2006).

[23] M. Djorić and M. Okumura, *Certain CR submanifolds of maximal CR dimension of complex space forms*, Differential Geom. Appl. **26/2**, 208–217, (2008).

[24] M. Djorić and M. Okumura, *Scalar curvature of CR submanifolds of maximal CR dimension of complex projective space*, Monatsh. Math. **154**, 11–17, (2008).

[25] M. Djorić and M. Okumura, *Certain condition on the second fundamental form of CR submanifolds of maximal CR dimension of complex projective space*, Israel J. Math. **169**, 47–59, (2009).

[26] M. Djorić and M. Okumura, *Certain condition on the second fundamental form of CR submanifolds of maximal CR dimension of complex hyperbolic space*, submitted.

[27] M. Djorić and M. Okumura, *Invariant submanifolds of real hypersurfaces of complex manifolds*, submitted.

[28] J. Erbacher, *Reduction of the codimension of an isometric immersion*, J. Differential Geom. **5**, 333–340, (1971).

[29]  A. Gray, *Principal curvature forms*, Duke Math. J. **36**, 33–42, (1969).

[30]  R. C. Gunning and H. Rossi, *Analytic functions of several complex variables*, Prentice-Hall, Englewood Cliffs, New Jersey, (1965).

[31]  R. Hermann, *Convexity and pseudoconvexity for complex submanifolds*, J. Math. Mech. **13**, 667–672, (1964).

[32]  M. Kimura, *Sectional curvatures of holomorphic planes on a real hypersurface in $P^n(C)$*, Math. Ann. **276**, 487–497, (1987).

[33]  S. Kobayashi and K. Nomizu, *Foundations of differential geometry II*, Interscience, New York, (1969).

[34]  M. Kon, *Pseudo-Einstein real hypersurfaces in complex space forms*, J. Differential Geom. **14**, 339–354, (1979).

[35]  S. G. Krantz, *Function theory of several complex variables*, AMS Chelsea Publishing, Providence, Rhode Island, (1992).

[36]  H. B. Lawson, Jr., *Rigidity theorems in rank-1 symmetric spaces*, J. Differential Geom. **4**, 349–357, (1970).

[37]  S. Montiel, *Real hypersurfaces of a complex hyperbolic space*, J. Math. Soc. Japan **37/3**, 515–535, (1985).

[38]  S. Montiel and A. Romero, *On some real hypersurfaces of a complex hyperbolic space*, Geom. Dedicata **20**, 245–261, (1986).

[39]  A. Newlander and L. Nierenberg, *Complex analytic coordinates in almost complex manifolds*, Ann. Math. **65**, 391–404, (1957).

[40]  R. Niebergall and P.J. Ryan, *Real hypersurfaces in complex space forms*, in *Tight and taut submanifolds* (eds. T.E. Cecil and S.-S. Chern), Math. Sci. Res. Inst. Publ. *32*, Cambridge University Press, Cambridge, 233–305, (1997).

[41]  R. Nirenberg and R.O. Wells, Jr., *Approximation theorems on differentiable submanifolds of a complex manifold*, Trans. Amer. Math. Soc. **142**, 15–35, (1965).

[42]  K. Nomizu, *Some results in E.Cartan's theory of isoparametric families of hypersurfaces*, Bull. Amer. Math. Soc. **79**, 1184–1188, (1973).

[43]  Z. Olszak, *Contact metric hypersurfaces in complex spaces*, Demonstratio Math. **16**, 95–102, (1983).

[44]  M. Okumura, *Certain almost contact hypersurfaces in Euclidean spaces*, Kōdai Math. Sem. Rep. **16**, 44–54, (1964).

[45]  M. Okumura, *On some real hypersurfaces of a complex projective space*, Trans. Amer. Math. Soc. **212**, 355–364, (1975).

[46]  M. Okumura, *Contact hypersurfaces in certain Kaehlerian manifolds*, Tôhoku Math. J. **18**, 74–102, (1966).

[47]  M. Okumura, *Submanifolds of real codimension of a complex projective space*, Acad. Nazionale dei Lincei **LVIII**, 544–555, (1975).

[48]  M. Okumura, *Codimension reduction problem for real submanifolds of complex projective space*, Colloq. Math. Soc. János Bolyai **56**, 574–585, (1989).

172    References

[49] M. Okumura and L. Vanhecke, *A class of normal almost contact CR–submanifolds in $\mathbb{C}^q$*, Rend. Sem. Mat. Univ. Politec. Torino **52**, 359–369, (1994).

[50] M. Okumura and L. Vanhecke, *n-Dimensional real submanifolds with $n - 1$-dimensional maximal holomorphic tangent subspace in complex projective space*, Rend. Circ. Mat. Palermo **XLIII**, 233–249, (1994).

[51] B. O'Neill, *The fundamental equations of a submersion*, Mich. Math. J. **18**, 459–469, (1966).

[52] G. de Rham, *Sur la réductibilité d'un espace de Riemann*, Comment. Math. Helv. **268**, 328–344, (1952).

[53] P. J. Ryan, *Homogeneity and some curvature conditions for hypersurfaces*, Tôhoku Math. J. **21**, 363–388, (1969).

[54] N. E. Steenrod, *The topology of fibre bundles*, Princeton University Press, Princeton, (1951).

[55] R. Takagi, *On homogeneous real hypersurfaces in a complex projective space*, Osaka J. Math. **10**, 495–506, (1973).

[56] R. Takagi, *Real hypersurfaces in a complex projective space with constant principal curvatures II*, J. Math. Soc. Japan **27**, 507–516, (1975).

[57] R. Takagi, *A class of hypersurfaces with constant principal curvatures in a sphere*, J. Differential Geom. **11**, 225–233, (1976).

[58] Y. Tashiro, *On contact structure of hypersurfaces in complex manifold I*, Tôhoku Math. J. **15**, 62–78, (1963).

[59] Y. Tashiro, *Relations between almost complex spaces and almost contact spaces*, Sûgaku **16**, 34–61, (1964), (in Japanese).

[60] Y. Tashiro and S. Tachibana, *On Fubinian and C-Fubinian manifolds*, Kōdai Math. Sem. Rep. **15**, 176–183, (1963).

[61] A.E. Tumanov, *The geometry of CR manifolds*, Encycl. of Math. Sci. 9 VI, Several complex variables III, Springer-Verlag, Berlin, 201–221, (1986).

[62] K. Yano, *Differential geometry on complex and almost complex spaces*, Pergamon Press, New york, (1965).

[63] K. Yano, *On harmonic and Killing vector fields*, Ann. Math. (2) **55**, 38–45, (1952).

[64] K. Yano, *Integral formulas in Riemannian geometry*, Marcel Dekker, New York, (1970).

[65] K. Yano and S. Bochner, *Curvature and Betti numbers*, Ann. Math. Studi. **32**, (1953).

[66] K. Yano and S. Ishihara, *Fibred spaces with invariant Riemannian metric*, Kōdai Math. Sem. Rep. **19**, 317–360, (1967).

[67] K. Yano and M. Kon, *CR submanifolds of Kaehlerian and Sasakian manifolds*, Progress in Mathematics, **30**, Birkhäuser, Boston, (1983).

[68] K. Yano and M. Kon, *Structures on manifolds*, World Scientific, Singapore, Series in Pure Mathematics, **3**, (1984).

# List of symbols

# Subject index